Investigations for a Course in Statistics

Martin A. Tanner
University of Wisconsin

MACMILLAN PUBLISHING COMPANY
New York

COLLIER MACMILLAN PUBLISHERS
London

MINITAB is a registered trademark.

Minitab, Inc.
3081 Enterprise Drive
State College, PA 16801 USA
Telephone: 814/238-3280
Fax: 814/238-4383

Printed in the United States of America

Macmillan Publishing Company
866 Third Avenue, New York, New York 10022

Collier Macmillan Canada, Inc.

Printing: 1 2 3 4 5 6 7 Year: 0 1 2 3 4 5 6

ISBN 0-02-418990-1

To Anat, Esther, Ida, Kalman and Noam

Preface

> There is something fascinating about science. One gets such wholesale returns of conjecture out of such a trifling investment of fact.
>
> -Mark Twain, *Life on the Mississippi*

Students who enroll in a course in statistics come with a wide range of interests, goals and expectations. In polling these students, I have found that the students, as a whole, respond with a common idea: Statistics somehow plays an important role in *doing* science and they have come to learn the details. The goal of this book is to illustrate the role of statistics in the process of *doing* science. While there are a number of excellent textbooks which cover such topics as elementary design, descriptive statistics and inference, there is no available means for the university student to *participate* in the act of investigation and to *experience* what statistics can and cannot do. The traditional statistical textbook is not directed toward creating such an *active* environment. Yet for the individual embarking on a career in science, it is important that they view statistics as a natural component of the dynamic process of investigation. To develop this outlook, they must see *statistics in action.*

The investigations included in this book highlight the interaction between science and statistics in several important ways. First, the students have the opportunity to act as scientist, as well as statistician. As a scientist, each student can observe and participate in the experimental process. This experience is invaluable in developing a sense of how the design of the experiment and the analysis of the data interact. As a statistician, each student can examine the data in a manner which will illustrate and solidify specific statistical concepts and methodology, as well as the general notion of chance variation. Moreover, as in any experimental setting, the investigation may be subject to such common problems as missing data and/or outliers. Clearly, a rigorous treatment of these issues is beyond the scope of a first course. However, it is important for the student to understand that ignoring the fact that some of the data are missing or ignoring outliers may lead to biased conclusions. It is important for the students to inquire why some observations could not be obtained. Are the data missing due to some "fluke" or are there additional variables which should be considered? In addition, it is also important for the students to inquire whether the outlying observations should be *downweighted* or *highlighted.* Investigations give the student the opportunity to work with data as the data would arise in their laboratory or in their database. Thus, the use of investigations in the introductory statistics course will enable students to see how real data really looks. (See Tanner, 1985.)

TO THE INSTRUCTOR:

The chapters included in this book share a common format. Each chapter begins with an INTRODUCTION that includes a list of key words and a description of the problem of interest. The chapters then continue with a survey of the necessary technical background required to analyze the data and a list of the material needed for the investigation. The purpose of the technical sections is to introduce a common notation and to briefly sketch the statistical methods and concepts which are highlighted in the investigation. The EXPERIMENTAL PROTOCOL section includes data sheets which are to be used to record the data. The data sheets can then be detached and handed in so that the class results may be accumulated and then returned to the students for analysis. The ANALYSIS section, which accompanies each experiment, provides a structured set of questions which serve as a guide to the student for an analysis of the data. As the questions tend to require careful consideration, it is recommended that the students work in teams with each student preparing his or her own write-up. The instructor is encouraged to supplement the analysis section with additional questions which highlight features of the particular data set. In addition, the instructor may wish to re-examine data sets as the course progresses and as new concepts are introduced. A list of references is given at the end of each investigation to direct the motivated student to fundamental papers in the literature. Each chapter concludes with a MINITAB APPENDIX to illustrate *how to do* the computations or graphs discussed in the body of the chapter. Again, it is recommended that the students tackle the computer work in teams so as to avoid getting hung-up with operational problems.

The investigations included in this book are designed so that they may be performed in the classroom, in the discussion section or at home. Suggestions for which mode is preferable for a particular experiment are given in the *Instructor's Notes* which accompanies the text. However, wherever the investigations are performed, it is important that the students *review* the experimental protocol before proceeding to its implementation. In this regard, the instructor or discussion section leader may wish to review the protocol with the students before its implementation. None of the investigations require special laboratory equipment to implement the activity. The results of the investigations can be examined in class or as a homework activity.

The chapters included in this book follow the general outline of a course in introductory statistics, as well as a course in statistical methods. The general topics addressed in the investigations include: elementary experimental design, descriptive and inferential statistics, elements of probability, discrete distributions, the normal distribution, one and two sample problems, regression, the analysis of categorical data, and sample surveys. In this way, these investigations are appropriate for use by sophomores to seniors, as well as beginning graduate students. This book was designed to be used as a supplement to a wide variety of textbooks, as illustrated in the following table:

CHAPTERS IN *Investigations for a Course in Statistics (ICS)* and other texts												
ICS	FPP	JB	BH	S	WW	O	MM	WH	J	MD8	MD9	DK
1	1, 2	1	1,6	1	1	1	4	1,15	1	1,9.1	1	1
2A	3,4	2	3	2, 3	2	2	1	2,3	2	2	2	2,3
2B	7,8	3	11.1	12	11,15	8.1	2,3	12	3	12	9	9
2C	26	10,11	11.1	9,10	9	4	7,8	9,10	8,10	7,8	6,7	7,8
3A	13	4	2	4	3	3	5	4	4	3	3	4
3B	28	14	2	14	17	6	3	11	11	11	8	11
4	5, 13-15	5-8	3,7	5-7	3,4,6	3	5,6	5-7	5-7	4-6	4,5	5,6
5	16-18	8	3	7	6,7	3	7	7	7	6	5	7
6	19-21	9	7	7,8	6,7	6	7	8,15	9	7	6	7
7	6,23,24	9	4	8	8	4	8	8	9	7	6	7
8	26	9-10	4	9,10	9	4	8	9	8	7	6	7
9	27	11	5	9,10	8,9	5	8	10	10	8,9	7	8
10A	28,29	14	8,9	14	17	6	9	11	11	11	8	11
10B	28,29	14	8,9	14	17	6	9	11	11	11	8	11
11	8-12	12-13	11	12	11,12	8,9	10	12	13	12	9	9

FPP - Freedman, Pisani and Purves (1978)
JB - Johnson and Bhattacharyya (1985)
BH - Brown and Hollander (1977)
S - Sincich (1990)
WW - Wonnacott and Wonnacott (1985)
O - Ott (1984)
MM - Moore and McCabe (1989)
WH - Weiss and Hassett (1987)
J - Johnson (1988)
MD8 - McClave and Dietrich (1988)
MD9 - McClave and Dietrich (1989)
DK - Dietrich and Kearns (1989)

A computer, whether micro, mini, supermini, or a mainframe, while not necessary, is recommended for use with this book. Though the size of the generated data sets will vary from class to class and from investigation to investigation, the students will certainly benefit from having access to a common data base (generated in the investigation) and by having the computer do the number crunching. These features, of course, allow the student to bypass the tedious entry of data, the tedious evaluation of formulas and to spend more time thinking about *what to do,* rather than *how to do it.* In addition, access to MINITAB which has graphic subroutines will give the student the flexibility to visually examine the data from a number of different perspectives. Whether a histogram, stem-and-leaf display or a scatter plot, graphs allow the eye to detect structure in the data and the ability to interpret graphical summaries of data comes with much practice. Each chapter in this book concludes with a MINITAB appendix to illustrate the how to do.

TO THE STUDENT:

Enjoy!

ACKNOWLEDGMENTS:

I wish to thank Robert L. Wardrop and Esther R. Tanner for their critical review of the manuscript and their very thoughtful comments.

Martin A. Tanner
Departments of Statistics
and Human Oncology
University of Wisconsin

REFERENCES

1. Brown, B. and Hollander, M. *Statistics: A Biomedical Introduction,* John Wiley, New York, 1977.

2. Dietrich, F. H. and Kearns, T. J. *Basic Statistics: An Inferential Approach,* Dellen Publishing Company, San Francisco, 1989.

3. Freedman, D., Pisani, R. and Purves, R. *Statistics,* W. W. Norton and Company, New York, 1978.

4. Johnson, R. *Elementary Statistics,* PWS-KENT Publishing Company, Boston, 1988.

5. Johnson, R. A. and Bhattacharyya, G. K. *Statistics: Principles and Methods,* John Wiley, New York, 1985.

6. McClave, J. T. and Dietrich, F. H. *Statistics,* Dellen Publishing Company, San Francisco, 1988.

7. McClave, J. T. and Dietrich, F. H. *A First Course in Statistics,* Dellen Publishing Company, San Francisco, 1989.

8. Moore, D. S. and McCabe, G. P. *Introduction to the Practice of Statistics,* W. H. Freeman and Company, New York, 1989.

9. Ott, L. *An Introduction to Statistical Methods and Data Analysis,* Duxbury Press, Boston, 1984.

10. Sincich, T. *Statistics by Example,* Dellen Publishing Company, San Francisco, 1990.

11. Tanner, M. A. "The Use of Investigations in the Introductory Statistics Course." *The American Statistician, 39,* 306-310, 1985.

12. Weiss, N. A. and Hassett, M. J. *Introductory Statistics,* Addison-Wesley Publishing Company, Reading, MA, 1987.

13. Wonnacott, R. and Wonnacott, T. *Introductory Statistics,* John Wiley, New York, 1985.

Contents

2C. The Shape of Things - Part C

3A. Are You Well Calibrated?

3B. The Vulcan Salute

4. Flip Out!

1

Counting Ease

Key Words: Statistics; Experiment; Randomization; Measurement Error; Stem-and-Leaf Plot; Outliers.

But time and chance happen to them all.
-The Book of Ecclesiastes

1.1 INTRODUCTION

This investigation will illustrate *statistics in action.* We will pose a question, perform an experiment to address this question, summarize the results and then draw a conclusion based on an examination of the data. In this way, the basic paradigm for scientific research will be illustrated. While the example is relatively simple, the paradigm that the example illustrates is used in a wide range of scientific activities. The question of interest that will be considered is whether the ability to recognize a letter of the English alphabet in any way depends on the letter. To address this question, some of you will be asked to count the number of times the letter "e" occurs in a paragraph, while others will be asked to count the occurrences of the letter "r". You will then examine the distributions of the students' errors for the "e" and for the "r" to assess whether there are any differences in the distributions of the measurement errors for the two letters.

1.2 THE AVERAGE AND STEM-AND-LEAF PLOT

1.2.1 THE AVERAGE

To analyze the data from the investigation, you will need to know how to compute the average and to draw a stem-and-leaf plot.

Given a list of n numbers, the **average** is the sum of the n numbers divided by n.

For example, the average of 1.2, 2.2, 3.2, 4.2 and 5.2 is

$$\frac{1.2 + 2.2 + 3.2 + 4.2 + 5.2}{5} = 3.2.$$

1.2.2 THE STEM-AND-LEAF PLOT

The best way to present the **stem-and-leaf plot** is by an example. A stem-and-leaf plot will be constructed for the following nineteen numbers: (The data represent self-measured pulse rates of statistics students at rest per one minute.)

68, 81, 80, 63, 59, 64, 48, 54, 72, 74, 75, 70, 55, 60, 67, 53, 64, 69, 75

The idea is to break each number into two components, the stem and the leaf. For these data, the tens' digit will form the stem and the ones' digit will form the leaf. The stems are plotted to the left of a vertical line, with each leaf plotted on the same line as its stem. Thus, the stem-and-leaf plot for the pulse data is given as:

```
4|8
5|3459
6|0344789
7|02455
8|01
```

For example, there is one number with a stem of 4, i.e. 48. The line with a stem of 5 has four leaves: 3, 4, 5 and 9. These values correspond to the values 53, 54, 55 and 59. The stem-and-leaf plot contains the same information as the list of numbers, but it presents the numbers in a graphical manner to show how the numbers are distributed. Examining the stem-and-leaf plot, it is seen that the pulse data seem to be symmetrically distributed about a central value. In fact, because the leaves are ordered within each row, it is easy to see that nine of the nineteen pulses are below 67 beats per minute and nine of the pulses are above 67 beats per minute. This stem-and-leaf plot does not contain any gaps (stems with no leaves), nor does it show any outliers (an **outlier** is an observation which is either much smaller or much larger than the rest of the data, i.e. a peculiar data point).

In general, each number is broken according to its leading digits and it trailing digits. Thus, when the numbers consist of three digits, one may break each number into its first two leading digits and its trailing digit. For example, 120 would become 12 and 0. Alternatively, one can break a three digit number into its first leading digit and its two trailing digits. For example, 120 would become 1 and 20. In such an instance, commas would be used to separate the stems.

There are no hard and fast rules for where to break a number. The only available strategy is to adopt the approach which leads to the more descriptive plot.

The stem-and-leaf plot (as well as other graphical techniques) is described in detail in Johnson and Bhattacharyya (1985). This technique is originally due to John Tukey who has developed a number of other ingenious techniques for data analysis. An introduction to his philosophy can be found in Tukey (1977).

1.3 MATERIALS

One coin (preferably new and of the same denomination) for each student and a copy of the text (see page 6).

1.4 EXPERIMENTAL PROTOCOL

NAME______________________

DATA SHEET

1. Record your height in **inches** ______________.

2. Record your weight in **pounds** ______________.

3. Record your GPA (A = 4.0) ______________.

4. Record your sex

   ______ **MALE** ______ **FEMALE**

5. How would you categorize your major? Check one of the following categories.

   1. Physical Science __________.
   2. Social Science __________.
   3. Biological Science __________.
   4. Humanities __________.
   5. Engineering __________.
   6. Business __________.
   7. Other or Undecided __________.

6. Do you require glasses or contact lenses for reading?

   ______ **YES** ______ **NO**

7. Flip the coin. Record the result.

   ______ **HEAD** ______ **TAIL**

8. If the toss resulted in a **HEAD,** then count the number of **e's** between the two black lines. (See the selection on the following page.) You are only allowed to count the e's once. Please skim the selection at the speed in which you would read a newspaper or magazine article. Do not collaborate with your friends.

My estimate of the number of e's between the two black lines is ______________.

If the toss resulted in a **TAIL,** then count the number of **r's** between the two black lines. (See the selection below.) You are only allowed to count the r's once. Please skim the selection at the speed in which you would read a newspaper or magazine article. Do not collaborate with your friends.

My estimate of the number of r's between the two black lines is ______________.

9. Hand in your responses to question 1-8, so that the class results can be pooled.

THE JOURNAL OF
NERVOUS AND MENTAL DISEASE

VOL. 170, NO. 8
August 1982
SERIAL NO. 1197

Operational Definitions of Schizophrenia

What Do They Identify?

MICHAEL A. YOUNG, PH.D.,[1] MARTIN A. TANNER, PH.D.,[2] AND HERBERT Y. MELTZER, M.D.[3]

Several studies have shown that different systems for diagnosing schizophrenia produce diagnoses with relatively low agreement. This, however, does not necessarily imply that the different systems are identifying conceptually different diagnostic groups. They may, in fact, identify a single entity, but with different degrees of accuracy. One-hundred ninety-six inpatients were diagnosed by multiple diagnostic systems. The pattern of relationship among the diagnoses was studied with latent class analysis, a statistical methodology which has not previously been applied to this type of data. Results indicated that the diagnoses of the Research Diagnostic Criteria, Flexible 6, Schneider, and 1978 Taylor and Abrams diagnostic systems all estimated a single underlying diagnosis. The Taylor and Abrams system identified this core diagnosis with high accuracy. The results also suggest that blunted affect and the absence of an affective syndrome are related to latent class schizophrenia, while the presence of only nonspecific delusions and/or hallucinations is not related. The appropriateness of the latent class definition of schizophrenia for selecting patients for research is discussed.

1.5 ANALYSIS

Your instructor will provide you with the class results, so that you may answer the following questions. In particular, you will be given a rectangular array of numbers with n rows, corresponding to the n students who participated in the investigation, and eight columns, corresponding to the responses to questions 1-8 of Section 1.4. For example, the first row will represent the responses of a student, where the number in the first column will be the student's height, the second number in the row will be the student's weight, and so on. Questions 1.1-1.9 examine whether the randomization led to two balanced groups. Questions 1.10-1.14 ask you to compare the distribution of the "e" and the "r" measurement errors, while the remaining questions relate to alternative designs for this investigation.

1.1. What percentage of the tosses resulted in a HEAD? How large a discrepancy from 50% would you have expected?

1.2. What percentage of the class counted the e's? Do your responses to 1.1 and 1.2 agree? (In examining data, it always pays to check the obvious.)

1.3. Of the people whose toss resulted in a HEAD, what percentage was male? Of the people whose toss resulted in a TAIL, what percentage was male? How do these two numbers compare?

1.4. Draw a stem-and-leaf plot of the GPA values for those people whose toss resulted in a HEAD. Discuss the shape of the plot. Draw a stem-and-leaf plot of the GPA values for those people whose toss resulted in a HEAD. Discuss the shape of the plot. Compare the two plots.

1.5. Compute the average weight of those people whose toss resulted in a HEAD. Compute the average weight of those people whose toss resulted in a TAIL. How do these two numbers compare? Compute the average height of those people whose toss resulted in a HEAD. Compute the average height of those people whose toss resulted in a TAIL. How do these two numbers compare? (If the HEAD and TAIL groups are balanced with respect to weight, would you expect these groups to be balanced with respect to height?)

1.6. Compare the HEAD and TAIL groups with respect to declared major. In particular, do each of these two groups have approximately the same proportion of physical science majors, social science majors, humanities majors, biological science majors and engineers?

1.7. Of the people whose toss resulted in a HEAD, what percentage requires glasses to read? Of the people whose toss resulted in a TAIL, what percentage requires glasses to read? Compare these two figures.

1.8. Based on your responses to questions 1.1 to 1.7, did the randomization seem to yield two balanced groups?

1.9. Are there other traits which may be of interest to examine for balance between the two groups?

1.10. Subtract the true number of e's (your instructor will know this number) between the two black lines from each of the "e" estimates to obtain the "e" **measurement errors.** Draw a stem-and-leaf plot for the "e" measurement errors.

1.11. Subtract the true number of r's (your instructor will know this number) between the two black lines from each of the "r" estimates to obtain the "r" **measurement errors.** Draw a stem-and-leaf plot for the "r" measurement errors.

1.12. Discuss the shape of each of the stem-and-leaf plots. How are the measurement errors distributed? Are the measurement errors symmetrically distributed about a particular value? Would you have expected the measurement errors to be symmetrically distributed about a particular value? Are there any gaps or peculiar points that show up in the plots?

1.13. Based on your responses to question 1.12, do the two distributions of the measurement errors differ? Does the general shape of the two distributions differ? If your answer is yes, then how do the shapes of the two distributions differ? (To compare the two stem-and-leaf plots, it may be helpful to place the two plots stem-to-stem.)

1.14. Based on the class results, does the ability to recognize an "r" differ from the ability to recognize an "e"? State a conclusion, based on your analysis of the data. What would you have expected?

1.15. Several alternative designs are available for this experiment. Critically discuss each of the following designs. What are their disadvantages? What are their advantages?

(a) Each student simultaneously counts the number of e's and r's .

(b) Each student first counts the e's in the paragraph, then the r's .

(c) Each student flips a coin. If the coin results in a HEAD, count the e's in the paragraph and then count the r's. If the coin toss results in a TAIL, count the r's in the paragraph and then count the e's. This design is called a *cross-over design* for the obvious reason that each subject crosses over from one treatment to the other treatment at some point in the study. (The book by

Friedman, Furberg and DeMets (1982) discusses this design in detail and provides further references.)

1.16. Ask your classmates what technique they used to enumerate the number of occurrences of the "e" or the "r" in the selection. List these techniques. Do you think that the technique of counting had an effect on the measurement error? How would you attempt to remove this effect?

1.17. Notice that you were asked to count the number of "e's" or the number of "r's" in the selection. If you were to take this request literally, you would have counted the number of occurrences of the three character symbol *e's* or the three character symbol *r's* in the selection. How would such an interpretation of the task have affected your result? Is it possible to determine which observations were collected in this manner? Would you include such observations in your analysis of the class results? Alternatively, you may have interpreted the request to count the number of "e's" or the number of "r's" in the selection as a request to enumerate the occurrences of the *lower case e* or the *lower case r*. How would such an interpretation of the task have affected your result? Is it possible to determine which observations were collected in this manner? Would you include such observations in your analysis of the class results? How would you redesign this experiment to eliminate the three character symbol and the lower case ambiguities?

1.6 MINITAB APPENDIX

The average of the numbers 2, 4, 6, 8 and 10 is computed by summing the numbers and then dividing this sum by the number of quantities in the list. As such, we can compute the sum 2 + 4 + 6 + 8 + 10 = 30 and then divide 30 by the number of quantities in the list (5), to obtain 30 / 5 = 6. How else can one compute the average of these five numbers? Some of you may have calculators with an $\overline{x}$ button, in which case to obtain the average, you first put your calculator into statistics mode, you then punch in the numbers 2, 4, 6, 8, and 10 and then hit the $\overline{x}$ button to obtain the number 6. If you had access to a computer, you could write a FORTRAN program or a PASCAL program to compute the average. Of course, writing a program to compute the average of five numbers is clearly not worth the effort. However, what if instead of five numbers you had five hundred numbers or five thousand numbers, would you still want to do the calculation by hand or even with a hand calculator?

A computer package called MINITAB is a collection of prewritten computer programs designed to perform some elementary and some not so elementary statistical computations. This package allows one to easily evaluate some very tedious and very time consuming formulas.

Had MINITAB only been a fancy number cruncher, it would not have become as popular as it has become. MINITAB has another feature which is really quite nice - graphics. With an appropriate command, MINITAB can be made to draw a stem-and-leaf plot. With other commands MINITAB can be made to implement a wide range of other graphical techniques which we will discuss in this book. Rather than spending minutes or even hours drawing a particular plot, MINITAB can do it in seconds. Such quick turn-around allows one to think about the data, rather than waste time and energy drawing the plot.

MINITAB works interactively. You, the user, begin a session by entering a command. MINITAB then looks up this command in its dictionary of commands and checks for errors. If MINITAB can recognize the command and there are no "fatal" errors in the command, then MINITAB will attempt to carry out the command. MINITAB will signal the completion of its implementation of the command with a prompt such as

MTB >

A command in MINITAB consists of a command name plus auxiliary information. The command name tells MINITAB what to do and the auxiliary information tells MINITAB on what to do it. To be more specific, MINITAB stores data in columns. For example, in response to the questions 1-4 in Section 1.4 everyone in the class should have recorded his or her height (in inches), weight (in pounds), GPA and sex. Thus, a set of numbers for a male student may be

72 173 3.52 0

(Notice that we have adopted the convention that male = 0 and female = 1.) Now, you may have forty students in your class. As such, you would have forty numbers in the height column, forty numbers in the weight column, forty numbers in the GPA column and forty numbers in the sex column. The way such data are stored in MINITAB is to put all the height data in a column (for example, column 1, C1), all the weight data in a column (for example, column 2, C2), all the GPA data in a column (for example, column 3, C3) and all the sex data in a column (for example, column 4, C4).

As an illustration, let us assume we had a class with five students, then our data may look like:

72 173 3.52 0
64 120 3.80 1
66 118 3.10 1
69 155 2.83 0
74 194 3.63 0

How would we get this information into MINITAB? We would first access MINITAB (your instructor will tell you how to do this since the chain of commands to access MINITAB varies between computers) and then type

READ the following data into columns C1-C4 *return*
72 173 3.52 0 *return*
64 120 3.80 1 *return*
66 118 3.10 1 *return*
69 155 2.83 0 *return*
74 194 3.63 0 *return*
END *return*

(The symbol *return* indicates that the return or enter button should be hit to signal the end of a line. We will, hereafter, suppress this symbol though it should be understood that hitting the return button is required at the end of each line.)

In general, MINITAB only needs the first four letters of a command and a column (or set of columns) which define the data upon which the command is applied. Thus, as far as MINITAB is concerned, to enter the above data you could have typed:

```
READ C1-C4
72 173 3.52 0
64 120 3.80 1
66 118 3.10 1
69 155 2.83 0
74 194 3.63 0
END
```

Either way, the numbers 72, 64, 66, 69 and 74 would be stored in column 1 (C1), the numbers 173, 120, 118, 155 and 194 in column 2 (C2) and so on. To print the data in columns C1, C2, C3 and C4, you can tell MINITAB to

PRINt the data in columns C1 C2 C3 and C4

or more simply PRIN C1-C4 . In addition, if you want to know the average height, you can type:

AVERage the data in column C1

or more simply AVER C1 , and MINITAB would reply with the answer of 69. (In general, MINITAB is not fussy regarding the use of upper and lower case letters. Upper case letters will be used to indicate the *required* components of a command and lower case letters to indicate the *optional* components.)

To complete the questions in Section 1.5, you will need to know how to get MINITAB to count the number of 1's in a column consisting of 1's and 0's. (For example, column C4.) The command SIGN C4 indicates how many 1's and 0's are in such a column. You will also need to know how to perform simple arithmetic computations using MINITAB. For example, to subtract the value 0.5 from each of the observations in column C1 and to then put the resulting differences in column C10, tell MINITAB to

SUBTract the value 0.5 from each of the observations in column C1
and put the differences in column C10

or more simply SUBT 0.5 C1 C10.

To have MINITAB draw a stem-and-leaf plot of the data in column C3, type

STEM-and-leaf of the data in column C3

or more simply STEM C3.

When you have completed your MINITAB session, you type STOP. After logging out of MINITAB you will need to log out of your computer account. Your instructor will have the details on how this is accomplished. As a final note, MINITAB has online documentation. For example, if you forget the format of a particular command or you want to do some statistical or arithmetic manipulation but you don't know the MINITAB command, try typing HELP and use the online documentation to solve your problem.

As wc discuss various graphical and numerical statistical techniques, we will introduce the appropriate MINITAB command to implement the particular technique. Ryan, Joiner and Ryan (1985) present a more complete exposition of the MINITAB system.

1.7 REFERENCES

1. Friedman, L., Furberg, C. and DeMets, D. *Fundamentals of Clinical Trials,* John Wright, Boston, 1982.

2. Johnson, R. A. and Bhattacharyya, G. *Statistics: Principles and Methods,* John Wiley, New York, 1985.

3. Ryan, B. F., Joiner, B. L. and Ryan, T. A. *Minitab Handbook,* Duxbury Press, Boston, 1985.

4. Tukey, J. W. *Exploratory Data Analysis,* Addison-Wesley, Reading, 1977.

2A

The Shape of Things That Are - Part A

Key Words: Bar Graph; Histogram; Shape; Location; Spread; Average; Median; Standard Deviation; Interquartile Range; Box-Plots.

There are three kinds of lies: lies, damned lies and statistics.
-Benjamin Disraeli

2A.1 INTRODUCTION

In this chapter, various graphical and numerical techniques will be presented for summarizing data. The bar graph, histogram and the box-plot will be presented. In addition, measures of location and measures of spread will be defined. The goal of this chapter's investigation will be to summarize the distribution of pocket change carried by the members of your class.

2A.2 GRAPHICAL SUMMARIES FOR UNIVARIATE DATA SETS

2A.2.1 BAR GRAPHS

One of the most widely used techniques for displaying qualitative data is the bar graph. The **bar graph** can display counts, such as how many observations fall into each of the categories of the qualitative variable. Or it can display proportions, such as what proportion of the observations fall into each of the categories of the qualitative variable. (It is up to you to decide whether to present the counts or the percents.)

To illustrate a use of the bar graph, consider the allocation of money by the National Endowment for the Arts in 1983 to its various programs (source: *Statistical Abstract of the United States: 1985,* 105th Edition). The National Endowment for the Arts supports several different programs: Music, State Art Agencies, Museums, Theater, Dance, Public Media, Challenge (a program designed to stimulate long-term sources of support for the arts), Visual Arts, and everything else called Other. Of the 145.3 million dollars awarded by the National Endowment for the Arts in 1983, the relative allocation according to program is given in the following table:

Federal Aid to the Arts

Program	Percent
State Arts	14.5
Music	8.9
Museums	6.9
Theater	6.5
Public Media	6.4
Dance	6.3
Visual Arts	4.0
Other	28.2
Challenge	18.3

The data, represented by a bar graph, are as follows:

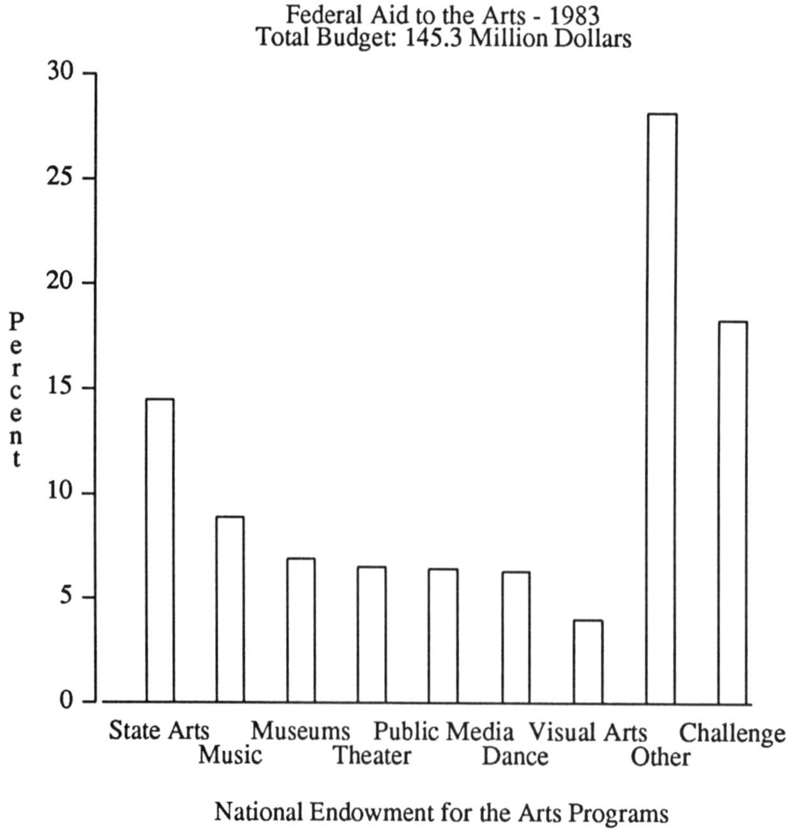

The most important thing to notice about the bar graph is that the height of each bar indicates the percentage of the budget awarded to the program represented by that bar; for example, just under 15% of the budget went to State Arts. This observation leads to the following rule:

A bar graph represents counts or percentages by the *height* of the bars, not by the area of the bars.

Because the bar graph conveys information by the height of each of the bars, the width of the bars are not varied. By varying both the width and the heights of the bars, the unwary reader may associate the percentage or count with the *area* of the bars instead of the height of the bars.

2A.2.2 HISTOGRAMS

The following data represent the weight in pounds of 42 students.

140, 103, 138, 140, 125, 145, 125, 143, 105, 125, 135, 140, 130, 150, 102, 165, 135, 175, 180, 153, 175, 165, 170, 140, 175, 185, 195, 183, 143, 185, 175, 215, 180, 185, 145, 150, 145, 207, 155, 165, 147, 180.

Let's look at a **histogram** for these data:

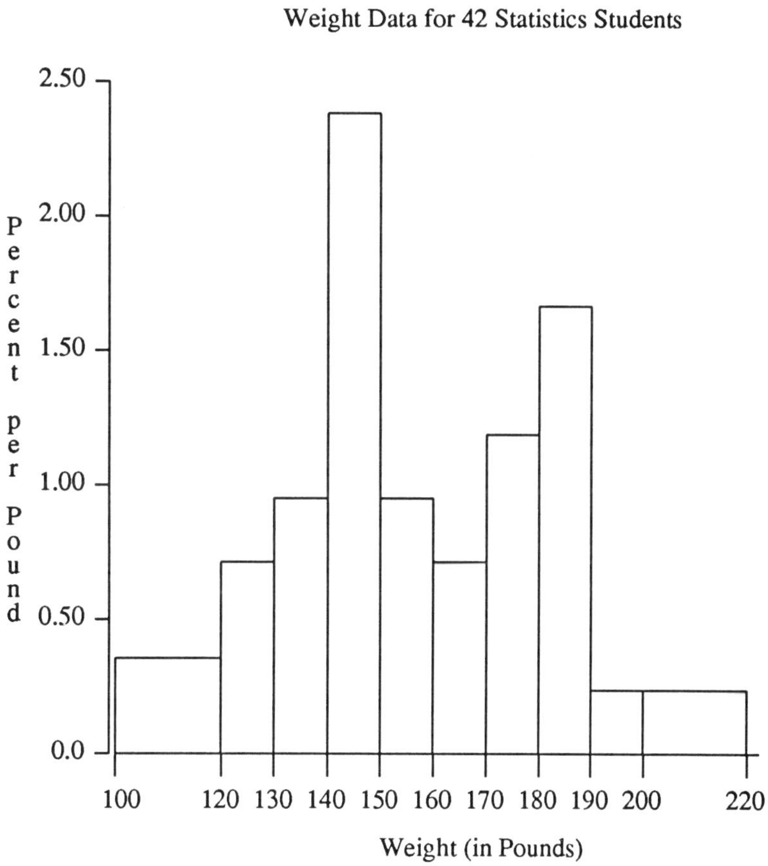

How does the histogram differ from a bar graph? First, note that the horizontal axis of the histogram represents a quantity, namely weight, which is measured on a numerical scale. In contrast, the horizontal axis of the bar graph represents a quality, namely the programs of the National Endowment for the Arts, which is categorical in nature. Next, note that the blocks of the histogram are not all the same width. Finally, note that the label on the vertical axis is not percent (or count) but rather percent per pound. Thus:

A histogram represents data by the *area* of the blocks, not by the height of the blocks.

To compute the height of the block over an interval, *one divides the corresponding percentage by the width of the interval.* In this way, the area of each block (height x width, i.e. percent per unit x unit) equals the percent and the sum of the areas of all the blocks equals 100%. For example, of the 42 statistics students, four reported weights between 130 and 139 pounds. In other words, 4 / 42 or 9.52% of the weights lie in the interval 130-139. Thus, the height of the block over this interval is 9.52% / 10 pounds = 0.952% / pound. If one were to calculate the area in this block, it would equal 9.52%. The sum of the areas of the ten blocks of this histogram is equal to 100%.

In general, one looks for three things in a histogram: **shape, location** and **spread.**

Shape indicates how the data are distributed. Are the data more or less symmetrically distributed about a point as:

Example of a Symmetric Histogram - Fake Data

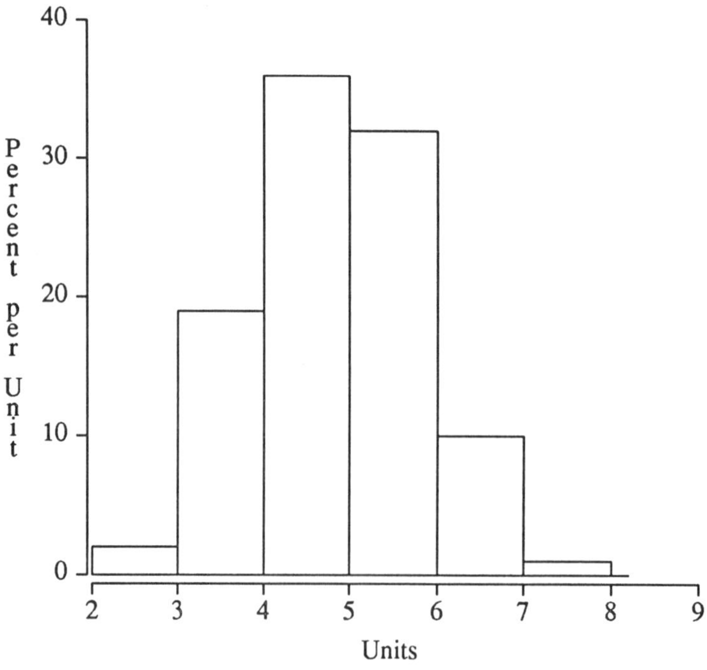

Or is the distribution of the data skewed as:

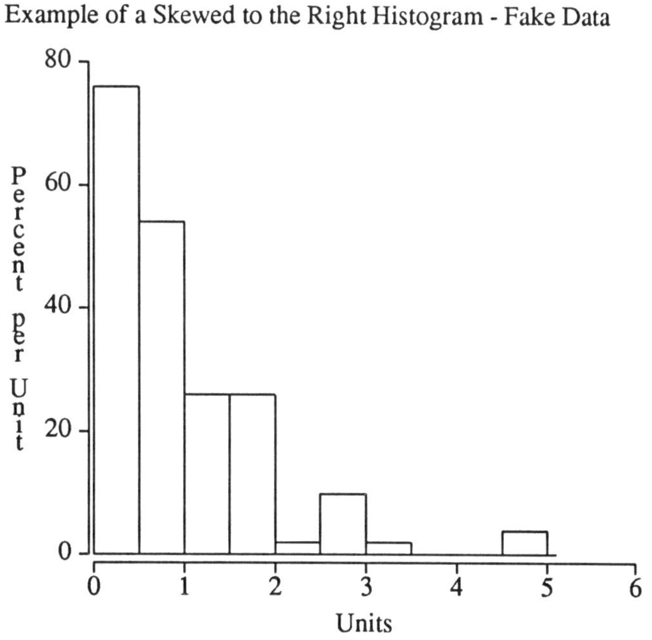

The above histogram is *skewed to the right* due to the presence of several high values in the data set. A histogram may be bimodal, that is, it may have two humps as does the histogram for our weight data.

Location indicates where the data tend to be located. For example, here are two more or less symmetric histograms which differ only in where the data tend to located.

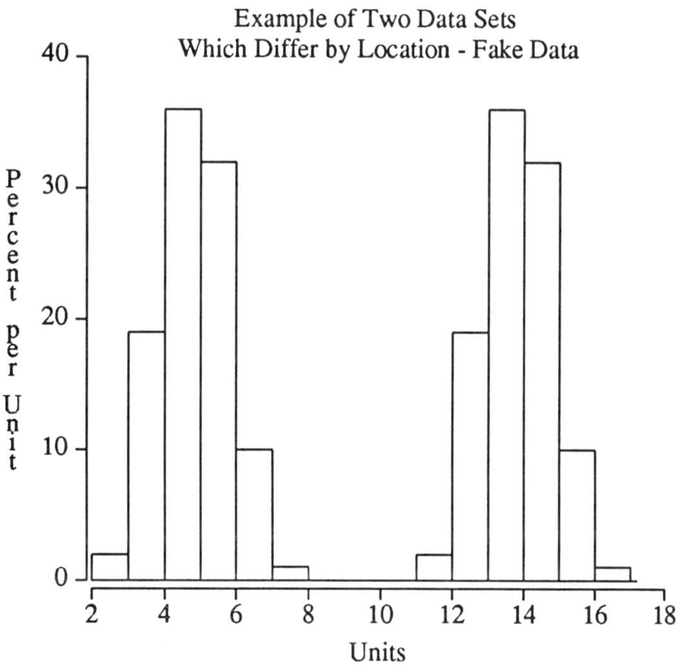

Spread indicates how dispersed the data tend to be. For example, here are two symmetric histograms which differ by location and by spread.

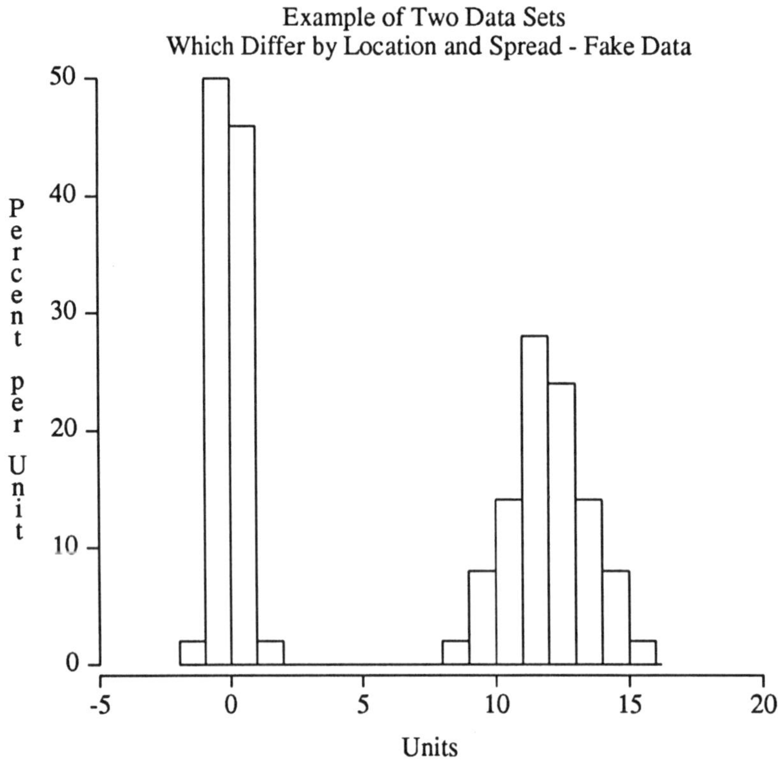

One of the more important problems considered in this book is the comparison of two or more data sets. For example, a comparison may be made between the distribution of the men's and women's heights. A good place to start in this comparison is a visual inspection of the data. Draw a histogram, draw a stem-and-leaf plot. In these plots, look for differences in *shape*. Watch for features in the plots such as spikes, gaps, symmetry, and skewness. Then, focus on issues of *location* and *spread*. To aid in the comparison of the plots, make sure that the axes of the plots are the same size and are drawn to the same scale. Put the graphs close enough to allow the reader to glance between the two figures.

2A.3 NUMERICAL SUMMARIES FOR UNIVARIATE DATA SETS

The histogram and stem-and-leaf plots are useful graphical techniques for summarizing quantitative data. They indicate how the data are distributed, where the data tend to lie and how spread out the data tend to be. When the histogram of a data set is symmetric and unimodal (one humped), it turns out that one only needs two numbers to describe the entire data set. These two numbers are: a measure of location and a measure of spread. The measure of location indicates where the histogram is centered, i.e. about what value the data are distributed. The measure of spread indicates how spread out the histogram is, i.e. how far the "typical" data point tends to be from the center point.

2A.3.1 MEASURES OF LOCATION

The average, which we will denote as $\bar{x}$, is probably the best known measure of location. (Note that some people use the word *mean* when referring to the average.)

To obtain the **average** of n numbers, sum the n numbers and divide by n.

The median is another measure of location. The **median** is the half-way point in the data. To be more precise:

If the number of observations is *odd,* then the median is the middle observation when the data are arranged in order.

If the number of observations is *even,* the median is the number halfway between the two middle observations when the data are arranged in order.

To illustrate the computation of the average and the median, consider the following seven reported grade point averages:

$$2.5, 2.8, 3.1, 3.3, 3.5, 3.6, 3.7\,.$$

Because the data are sorted, it can be immediately seen that the median for these seven observations is 3.3. If these seven numbers are summed and then divided by seven, it is found that the average is 3.21. In other words,

$$3.21 = \frac{2.5 + 2.8 + 3.1 + 3.3 + 3.5 + 3.6 + 3.7}{7}\,.$$

2A.3.2 MEASURES OF SPREAD

To assess how far, on average, the numbers in a list tend to be from the center of the data, we will require a measure of spread. A very popular measure of spread is the standard deviation, which we will denote as *SD*. (Note that some people use *sd,* to denote the standard deviation, while other people use *s*. We will use the notation *SD* in this book.) The **standard deviation** corresponding to the numbers x_1, x_2, ... and x_n, is given by the formula:

$$SD = \sqrt{\frac{(x_1 - \bar{x})^2 + (x_2 - \bar{x})^2 + \cdots + (x_n - \bar{x})^2}{n-1}}\,,$$

where $\bar{x}$ is the average of the numbers in the list. It is important to note that the standard deviation has the same units as the data. That is, if the data are measured in feet, then the

SD will be in feet; if the data are in kilograms, then the *SD* will be in kilograms. By "averaging" the squared distances of the numbers from the average, *the standard deviation has the interpretation of measuring the typical distance of a number in the list from the average* ($\overline{x}$) *of the list.*

How does the *SD* help describe a list of numbers? For a list of numbers which yield a more or less bell-shaped histogram, about seventy percent of the numbers will fall between $\overline{x}$ - *SD* and $\overline{x}$ + *SD*. In other words, about seventy percent of the numbers will fall within one standard deviation of the average. In addition, about ninety-five percent of the numbers will fall between $\overline{x}$ - 2*SD* and $\overline{x}$ + 2*SD*. That is, about ninety-five percent of the numbers will fall within two standard deviations of the average. Given the average, the *SD* and the fact that the histogram of the list of numbers is more or less bell-shaped, one can construct the range within which lies 70% of the numbers and the range within which lies 95% of the numbers.

An alternative measure of spread, the interquartile range, measures the span of the central 50% of the numbers in a list of numbers. Since the interquartile range measures the span of the central numbers, it is less sensitive to outlying values than the standard deviation. To define the interquartile range, two numbers are introduced: Q_1 and Q_3. Q_1, the first quartile of the distribution, is a number at or below which at least 25 percent of the observations are found, and at or above which at least 75 percent are found. Q_3, the third quartile of the distribution, is a number at or below which at least 75 percent of the observations are found, and at or above which at least 25 percent are found.

Note that two adjacent values may satisfy the definition of Q_1 and Q_3. For example, in the data set

2.5, 2.8, 3.1, 3.3, 3.5, 3.6, 3.7, 3.8,

both 2.8 and 3.1 satisfy the definition for Q_1 and both 3.6 and 3.7 satisfy the definition for Q_3. In such a situation, we will adopt the convention that the quartile is the average of the two adjacent values. (Other books may adopt a different convention.) Thus, in this case, $Q_1 = 2.95$ and $Q_3 = 3.65$.

Having defined the first and third quartile, Q_1 and Q_3, we have:

The **interquartile range** equals $Q_3 - Q_1$.

In this way, the interquartile range represents the range within which lies the central half of the numbers in the list.

As an example of the computation of the interquartile range and the standard deviation, let's reconsider the seven grade point averages:

2.5, 2.8, 3.1, 3.3, 3.5, 3.6, 3.7 .

Q_1 equals 2.8. It is seen that at least twenty-five percent (actually 28.6%, since 2 / 7 = 0.286) of the numbers are at or below 2.8 and at least seventy-five percent (actually 85.7%, since 6 / 7 = 0.857) of the observations are at or above 2.8. Q_3 equals 3.6.

Moreover, it is seen that at least seventy-five percent (actually 85.7%, since 6 / 7 = 0.857) of the observations are at or below 3.6 and at least twenty-five percent (actually 28.6%, since 2 / 7 = 0.286) of the observations are at or above 3.6. For this list of numbers, the interquartile range is then 3.6 - 2.8 = 0.8. In other words, the range of the central reported grade point averages is 0.8.

To compute the standard deviation, one must evaluate the formula

$$\sqrt{\frac{(2.5-3.21)^2+(2.8-3.21)^2+(3.1-3.21)^2+(3.3-3.21)^2+(3.5-3.21)^2+(3.6-3.21)^2+(3.7-3.21)^2}{6}},$$

where 3.21 is the average of the seven numbers. Working through the arithmetic, it is found that the standard deviation of the grade point averages is about 0.44. Note that for this list of numbers, 71% (5 / 7) of the grade point averages are within one standard deviation of the average, that is, between 2.77 (= 3.21 - 0.44) and 3.65 (= 3.21 + 0.44). Moreover, 100% of the numbers on the list are within two standard deviations of the average, that is, between 2.33 (= 3.21 - 0.88) and 4.09 (= 3.21 + 0.88).

2A.4 THE BOX-PLOT

The **box-plot** is a graphical technique that shows both the center and spread of the data. It is less refined than the stem-and-leaf plot, since it cannot be used to recover the individual observations. It is also less refined than the histogram, since such features as bumps or gaps are hidden in the box-plot. The role of the box-plot is to give a quick impression of the data. The box of the box-plot covers the range of the central 50% of the observations. The antenna which emerges from the top of the box indicates how the data in the upper tail are distributed, the foot indicates how the data in the lower tail are distributed and any stars in the plot highlight outlying points.

To illustrate the box-plot, consider the data collected when students were asked to record the amount of change, in cents, (coins, not paper money) in their pockets, wallets, purses, socks and so on. The data expressed in cents, are as follows:

2, 126, 15, 78, 135, 0, 90, 0, 242, 235, 301, 146, 142, 79, 87, 7, 24,
190, 17, 60, 40, 37, 35, 0, 30, 145, 77, 0, 27, 89, 0, 74, 0, 87, 40,
71, 30, 2, 0, 0, 66, 50, 0, 126, 65, 99, 30, 95, 199, 8, 65 .

Here is a box-plot of the resulting data:

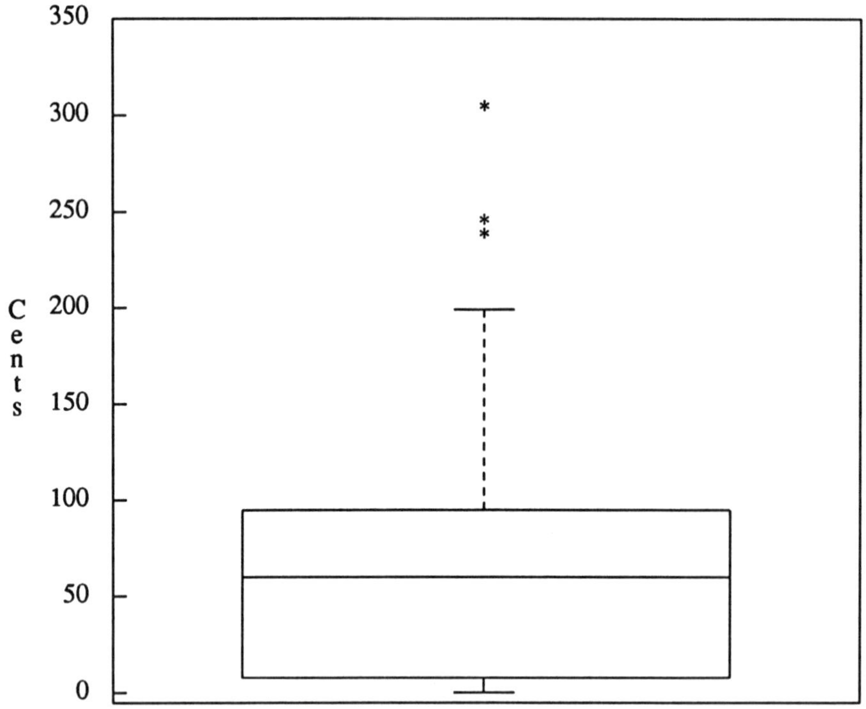

Distribution of Change

The upper and lower quartiles (Q_1 and Q_3) of the data are represented in the box plot by the top and bottom of the box. In this case, one can see from the plot that Q_1 is about 10 cents (actually 8) and Q_3 is a little less than 100 cents (actually 95). The median is represented by the horizontal line within the box, which in this case is 60 cents.

What is that antenna sticking out of the top of the box? The length of the antenna sticking out of the top of the box is about 1.5 times the interquartile range (IQR). The word "about" is used because the top antenna runs from the top of the box (Q_3) to the largest observation which is less than the value Q_3 + 1.5 x IQR. For example, in our change data set, Q_3 is equal to 95, while Q_1 is equal to 8. In this way, the IQR is equal to 87 and the product 1.5 x IQR is equal to 130.5. If we then compute Q_3 + 1.5 x IQR, we find this value is equal to 225.5. Thus, the antenna is drawn from the top of the box to the largest observation which is smaller than 225.5. In the present case, this value is 190 cents.

What is that foot sticking out of the bottom of the box? In general, the foot of the box-plot runs from the bottom of the box to the smallest observation which is greater than Q_1 - 1.5 x IQR. For our change data, the foot runs from 8 cents down to zero cents. Note that Q_1 - 1.5 x IQR is equal to -122.5. The smallest observation in the list, which is greater than -122.5, is 0 cents.

What are those stars? Any observation which is greater than Q_3 + 1.5 x IQR or is less than Q_1 - 1.5 x IQR is plotted as an individual point. For our change data, there are three such points.

One feature which comes across quite clearly in this box-plot is that these data are highly asymmetric. Note that the median is, more or less, centered within the box. Thus, it appears from the box-plot, that the central 50% of the data are symmetrically distributed about the median. However, note that the antenna is not the same length as the foot. That is, observations in the upper tail of the distribution, i.e. the observations larger than Q_3 tend to be distributed over a larger range than the observations in the lower tail of the distribution.

2A.5 MATERIALS

Each student should have the data sheet.

2A.6 EXPERIMENTAL PROTOCOL NAME____________________

DATA SHEET

Please answer the following questions:

1. How much change **(in cents)** do you have on your person? ________________

   (Total the value of the coins, not the paper money.)

2. I am

   _________Male _________Female

3. Record your GPA (A = 4.0)______________.

4. How would you categorize your major? Check **one** of the following categories.

   1. Physical Science __________.
   2. Social Science __________.
   3. Biological Science __________.
   4. Humanities __________.
   5. Engineering __________.
   6. Business __________.
   7. Other or Undecided __________.

Hand in this page so that the data may be pooled.

2A.7 ANALYSIS

Your instructor will provide you with the results of the investigation, so that you will be able to answer the following questions: In questions 2A.1 and 2A.2, you will examine the change data. You will examine the shape of the distribution of these data, their location, their spread and whether there are any outlying values. Questions 2A.3-2A.6 explore the relation between sex and change, while 2A.7 and 2A.8 explore the relation between GPA and change and academic major and change, respectively.

2A.1. Draw a stem-and-leaf plot for all the change data. Describe the distribution of the data. (Remember Shape, Location and Spread!) Is the distribution symmetric or skewed? Are there any outliers in the data? Are there any peculiarities in the data?

2A.2. Compute the average and median amount of change. Given the shape of your stem-and-leaf plot, would you have expected these numbers to be similar? Compute the standard deviation of the data. What percentage of the data are further than two standard from the average? Compute the interquartile range of the data. Over what range do the central fifty percent of the data span?

2A.3. Draw a stem-and-leaf plot for the men's change. Draw a stem-and-leaf plot for the women's change. What do you see (remember SLAS!) in each stem-and-leaf plot? Compare the two graphs. In what ways are the graphs similar? In what ways do the graphs differ?

2A.4. Divide each of the women's observations by 2 and repeat question 2A.3. State a conclusion regarding how men and women differ in the amount of change they carry on their person.

2A.5. Draw a box-plot for the men's change. Draw a box-plot for the women's change. Describe the two box-plots. Compare the two plots. How are they similar? How are they different?

2A.6. Divide each of the women's observations by 2 and repeat question 2A.5. State a conclusion regarding how men and women differ in the amount of change they carry on their person.

2A.7. Categorize each person as a "Whiz" or a "non-Whiz" according to whether or not their GPA is above 3.25. Compare these two groups, as you did men and women in questions 2A.3 and 2A.5.

2A.8. Compare the academic majors with regard to the distribution of change using stem-and-leaf plots and box-plots. Describe the stem-and-leaf plots. Compare the plots. How are they similar? How are they different? Describe the box-plots. Compare the plots. How are they similar? How are they different? State a conclusion regarding how the academic majors differ in the amount of change they carry on their person.

2A.8 MINITAB APPENDIX

If the weights of the forty-two statistics students were stored in column C1 and MINITAB was given the command

HISTogram of the data in column C1

or more simply HIST C1 , then MINITAB would respond with

```
histogram of C1    N = 42

Midpoint    Count
     100        2   **
     110        1   *
     120        0
     130        4   ****
     140        9   *********
     150        7   *******
     160        1   *
     170        4   ****
     180        8   ********
     190        3   ***
     200        1   *
     210        1   *
     220        1   *
```

As you can see, the MINITAB histogram is somewhat different from the histogram presented in this chapter. MINITAB displays the histogram on its side and does not label the axis as percent per unit. In addition, MINITAB displays on the plot how many weights fall in a particular interval. For example, in the interval centered at 130 (running from 125 to 134) there are four values.

In MINITAB, there is some flexibility as to the specification of the histogram. For example, the command

HISTogram of the data in column C1 with the first interval centered at 105 and all intervals of width 10

or more simply HIST C1 105 10 will yield a histogram whose first interval is centered at 105 with all intervals of width 10

```
Histogram of C1   N = 42

Midpoint   Count
   105.0       3  ***
   115.0       0
   125.0       3  ***
   135.0       4  ****
   145.0      10  ***********
   155.0       4  ****
   165.0       3  ***
   175.0       5  *****
   185.0       7  *******
   195.0       1  *
   205.0       1  *
   215.0       1  *
```

The DESCRIBE commands computes several common summary statistics for a list of numbers. For example, given the list of numbers 2.5, 2.8, 3.1, 3.3, 3.5, 3.6, 3.7 which were stored in column C1, the command

DESCribe the data in column C1

or more simply DESC C1 yields

```
              N      MEAN   MEDIAN   TRMEAN    STDEV   SEMEAN
C1            7     3.214   3.300    3.214     0.441   0.167

            MIN       MAX       Q1       Q3
C1        2.500     3.700    2.800    3.600
```

For these data, the average is 3.214, the median is 3.3, the standard deviation is 0.441, the minimum observation is 2.5, the maximum observation is 3.7, the first quartile is 2.8 and the third quartile is 3.6. (Note that when two adjacent numbers satisfy the definition for Q_1 or Q_3, then MINITAB adopts a different approach than presented in this book.) The quantity TRMEAN will not be discussed in this book, so you can ignore it. Suffice it to say, if the distribution of the data is symmetric, then the TRMEAN will be close to the average. We will return to the SEMEAN later in this book.

How did I get the numbers 2.5, 2.8, 3.1, 3.3, 3.5, 3.6, 3.7 into MINITAB? The SET command was used. In particular, I typed

SET the following data into column C1
2.5, 2.8, 3.1, 3.3, 3.5, 3.6, 3.7
END

To obtain a box-plot of the data in column C1, just tell MINITAB

BOXPlot of the data in column C1

or more simply BOXP C1.

You will also need to know how to select the women's weights and how to select the men's weights from the weights of all the students in the class. To accomplish this task, use the COPY command. In particular, if the following data were stored in columns C1 and C2, respectively, (men = 2 and women = 1),

134 1
133 1
145 1
186 2
199 2
102 1
155 2

then the command sequence

COPY the data from C1 to C3;
USE only the numbers whose partner in C2 is equal to 2.

or more simply,

COPY C1 C3;
USE C2 2.

will set up column C3 with the data

186
199
155

In other words, column C3 will contain the men's weights. The command sequence

COPY C1 C4;
USE C2 1.

will put the women's weights in column C4.

Finally, to sort the data in a column, use the command

SORT the data in column C1 and put the sorted values in column C2

or more simply SORT C1 C2. This command followed by PRIN C2 will print the data sorted from smallest to largest.

2B

The Shape of Things That Are - Part B

Key Words: Scatter Plot; Correlation Coefficient.

> That is like assuring the non-swimmer that he can safely walk across a river because its average depth is only four feet.
> -Milton Friedman

2B.1 INTRODUCTION

In this investigation, the data collected in the investigation for Chapter 1 will be examined. Before proceeding with the analysis, the scatter plot and correlation coefficient will be introduced.

2B.2 A GRAPHICAL SUMMARY FOR BIVARIATE DATA SETS

The most powerful graphical technique for analyzing how two variables relate is the scatter plot. Suppose one has n pairs of numbers, (x_1,y_1), (x_2,y_2) ... (x_n, y_n), where, for example, x_i represents the height of the ith person in the class and y_i represents the corresponding weight. The **scatter plot** of weight versus height is obtained by plotting y_i against x_i.

The following table presents forty-two pairs of heights and weights:

Height	*Weight*	*Height*	*Weight*
65	140	75	165
63	103	73	170
66	138	65	140
68	140	72	175
63	125	72	185
65	145	76	195
64	125	61	183
65	143	61	143
63	105	72	185
64	125	71	175
67	135	76	215
67	140	71	180
67	130	76	185
71	150	67	145
68	102	68	150
74	165	69	145
66	135	72	207
74	175	71	155
71	180	68	165
71	153	69	147
69	175	70	180

For these data, the following scatter plot was obtained:

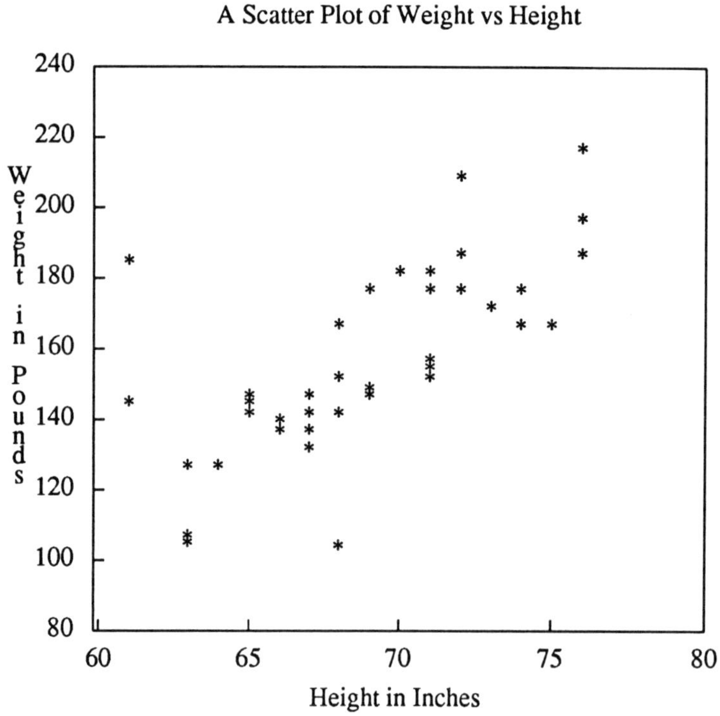

In *general,* one observes an upward trend to the plot, that is, taller people tend to be heavier, while shorter people tend to be lighter. In other words, there appears to be an *association* between height and weight. In fact, the association is *positive,* as tall people are heavier and short people are lighter.

Some of you may have questioned why the scatter plot is pooled over men and women, thus ignoring a feature of the data. To aid our analysis of the data, it will be useful to display the sex information. One simple approach would be to label those points corresponding to a woman as "w" and those corresponding to a man as "m". In this way, observe:

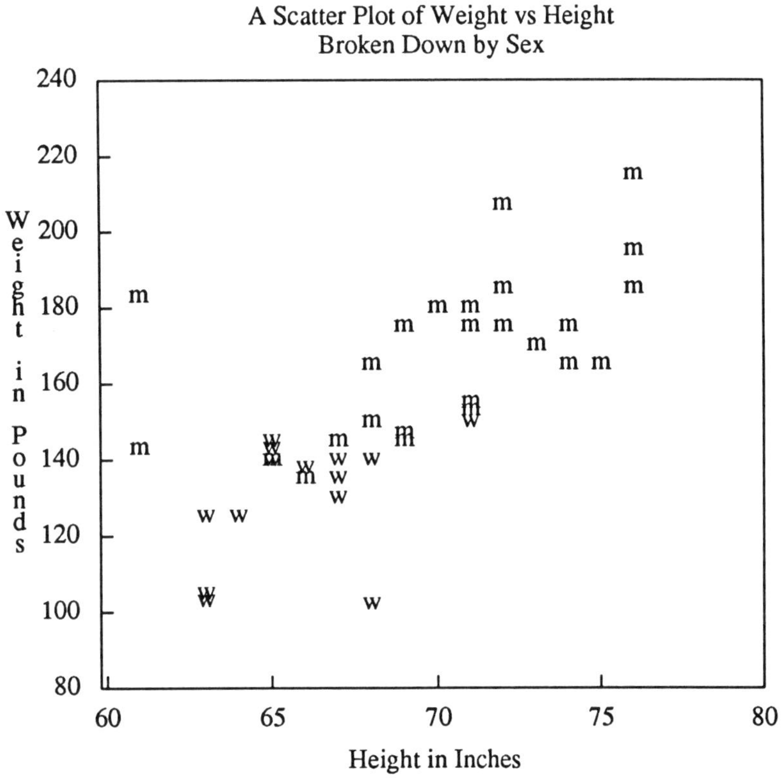

From this plot, it can be seen that the women are lighter and shorter than the men, *on average.* The men tend to cluster in the upper right of the plot, while the women tend to cluster in the bottom left of the plot. However, there are men who are shorter than and lighter than some of the women.

2B.3 A NUMERICAL SUMMARY FOR BIVARIATE DATA

The correlation coefficient is a measure of the association between two quantitative variables. Given the n pairs of numbers (x_1,y_1), (x_2,y_2), ... (x_n,y_n), the **correlation coefficient** (r) is given by:

$$r = \frac{[\,(x_1 - \overline{x})(y_1 - \overline{y}) + (x_2 - \overline{x})(y_2 - \overline{y}) + \cdots + (x_n - \overline{x})(y_n - \overline{y})\,] \;/\; n}{SD_x^{\,+}\; SD_y^{\,+}} ,$$

where

$$SD_x^{\,+} = \sqrt{\frac{(x_1 - \overline{x})^2 + (x_2 - \overline{x})^2 + \cdots + (x_n - \overline{x})^2}{n}} ,$$

and

$$SD_y^{\,+} = \sqrt{\frac{(y_1 - \overline{y})^2 + (y_2 - \overline{y})^2 + \cdots + (y_n - \overline{y})^2}{n}} .$$

Thus, to compute the correlation coefficient, one averages the $(x_i - \overline{x})(y_i - \overline{y})$ terms and then divides by the modified standard deviation of the x_i's and the modified standard deviation of the y_i's.

Note that if both numbers of a given pair (x_i, y_i) are above their respective average ($\overline{x}$ and $\overline{y}$), then the product $(x_i - \overline{x})(y_i - \overline{y})$ will be greater than zero. If both numbers of the pair (x_i, y_i) are below their respective average, then the product $(x_i - \overline{x})(y_i - \overline{y})$ will again be greater than zero. However, if exactly one of the numbers of the pair is less than its average, then the product $(x_i - \overline{x})(y_i - \overline{y})$ will be less than zero. Now notice that the correlation coefficient formula says to average all such products. In this way, if the scatter plot indicates a positive association, i.e. the terms $(x_i - \overline{x})(y_i - \overline{y})$ tend to be positive in value, then the correlation coefficient will be greater than zero. On the other hand, if the data indicate a negative association, i.e. the terms $(x_i - \overline{x})(y_i - \overline{y})$ tend to be negative in value, then the correlation coefficient will be less than zero.

The correlation coefficient measures how tightly the scatter cloud of points clusters *about a line*. When the correlation coefficient is equal to +1.0, all the points in the cloud fall on a line which slopes *up*. As the following figures indicate, as the correlation coefficient increases from 0.0 to 1.0, the scatter about the line is seen to *decrease*. An r equal to 1.0 corresponds to all the points falling on the line. An r equal to 0.0 corresponds to random scatter.

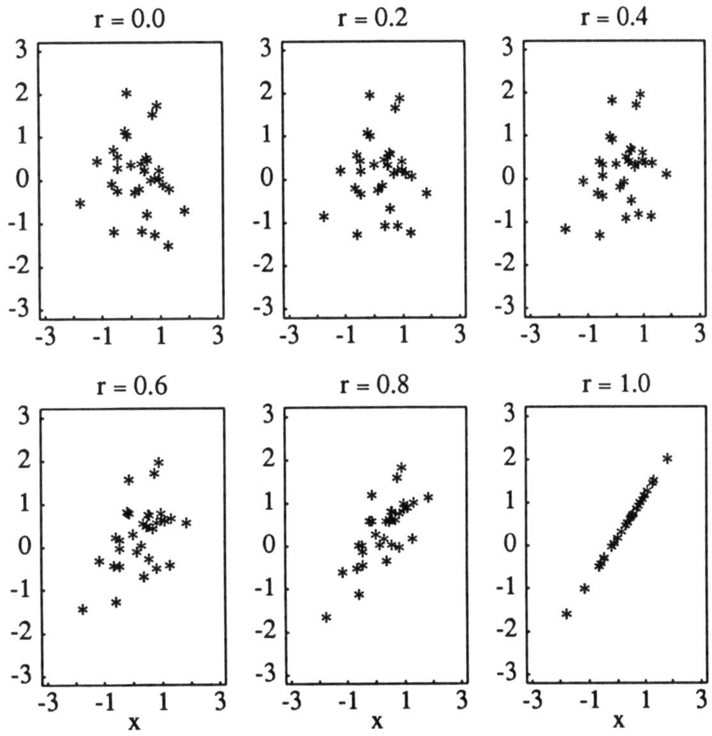

When the correlation coefficient is equal to -1.0, all the points in the cloud fall on a line which slopes *down*. As the correlation coefficient changes from -1.0 to 0.0, the scatter about the line is seen to *increase*.

The number r^2 quantifies the proportion of variability in one variable which is explained by knowledge of the second variable and vice versa. For example, consider the panel above where $r^2 = 1.0$. Note that the y variable ranges from -2 to 2. However in that panel, given a value for x, one can determine y exactly since all the points fall on a straight line. In other words, the linear relationship on x explains 100% of the variation of y. In the lower left panel ($r^2 = 0.36$), the y variable also ranges from -2 to 2. In that case, for a given value of x, one cannot predict y exactly. In that case, the association between x and y accounts for a fraction of the variation in y. In fact, because $r^2 = 0.36$, one can say that knowledge of x explains 36% of the variation of the y's.

For the height and weight data of Section 2B.2, r is equal to 0.71. In this way, r^2 is equal to $0.71^2 = 0.50$. Thus, one can say that fifty percent of the variability in height can be explained by weight and fifty percent of the variability in weight can be explained by height.

In summary,

a. The correlation coefficient is a measure of association between two variables. The value is positive when the association is positive and is negative when the association is negative.

b. The correlation coefficient measures linear association, i.e. how tightly the points in the scatter plot cluster about a straight line.

c. The correlation coefficient takes values between -1 and +1 and is a unitless number. When the correlation coefficient equals +1, all the points fall on a line with a positive slope. When the correlation coefficient equals -1, all the points fall on a line with a negative slope. A correlation of zero may indicate random scatter.

d. The square of the correlation coefficient measures the proportion of variability of one variable which is explained by the other variable.

2B.4 MATERIALS

The results of the investigation in Chapter 1.

2B.5 ANALYSIS

Your instructor will provide you with the results of the investigation in Chapter 1 so that you will be able to answer the following questions: In questions 2B.1-2B.6, you will examine the weight data. You will examine the shape of the distribution of these data, their location, their spread and whether there are any outlying values. In question 2B.7, you will be asked to repeat this analysis for the height data. Questions 2B.8 and 2B.9 relate to the GPA and academic major data, while question 2B.10 explores a possible relation between these two variables. Question 2B.11 explores the relation between height, weight and sex, while questions 2B.12 and 2B.13 explore the relation between height, weight and vision and the relation between height, weight and GPA, respectively.

2B.1. Draw a stem-and-leaf plot for all the weights. Describe the distribution of the weights. (Remember Shape, Location and Spread! - SLAS) Is the distribution symmetric or skewed? Are there any outliers in the data? Are there any peculiarities to the reported weights?

2B.2. Compute the average weight and the median weight. Given the shape of your stem-and-leaf plot, would you have expected these numbers to be similar? Compute the standard deviation (*SD*) of the weights. What percentage of the weights are further than two standard deviations from the average? Compute the interquartile range of the data. What percentage of the weights are further than one interquartile range from the median?

2B.3. Draw a stem-and-leaf plot for the men's weights. Draw a stem-and-leaf plot for the women's weights. What do you see (remember SLAS!) in each stem-and-leaf plot? Compare the two graphs. In what ways are the graphs similar? In what ways do the graphs differ?

2B.4. Compute the average of the men's weights and the average of the women's weights. Compute the median of the men's weights and the median of the women's weights. How much heavier is the typical man than the typical woman?

2B.5. Are there any *outliers* among the men's weights? Are there any *outliers* among the women's weights? If yes, can you explain why these values are much smaller or much larger than the rest of the data? If you were to delete these outlying observations, how would your responses to question 2B.4 change?

2B.6. Draw a box-plot for the men's weights. Draw a box-plot for the women's weights. Describe the two box-plots. Compare the two plots. How are they similar? How are they different?

2B.7. Repeat questions 2B.1-2B.6 for the height data.

2B.8. Draw a separate stem-and-leaf plot for the men's and women's GPA data. What do you see in each plot? Compare the two plots. How are the plots similar? How do they differ? Compare the women's average GPA to the men's average GPA. Compare the women's median GPA to the men's median GPA. How does the typical man compare to the typical woman with regard to GPA?

2B.9. Draw a bar graph for the academic major data. Describe this graph.

2B.10. Draw a bar graph of the academic major data for those people with reported grade point averages greater than or equal to 3.0. Draw a bar graph of the academic major data for those people with a reported grade point average below 3.0. Describe each bar graph. Compare these bar graphs. Does the reported grade point average appear to be related to the choice of major?

2B.11. (a) Draw a plot of weight versus height. Do the variables appear to be associated? Compute the correlation coefficient and interpret r^2.

(b) Indicate on the plot which points correspond to women and which to men. What do you see? Does a student's height and weight appear to be related to the student's sex? Compute the correlation coefficient and interpret r^2 for the men. Compute the correlation coefficient and interpret r^2 for the women. Contrast the association of height and weight for the men to the corresponding association for the women.

2B.12. Indicate on the height-weight plot which points correspond to contact lens or eye glass wearers. What do you see? Does a student's height and weight appear to be related to the student's vision? Compute the correlation coefficient and interpret r^2 for the contact lens / eye glass people. Compute the correlation coefficient and interpret r^2 for the other people in the class. Contrast the association of height and weight for the contact lens / eye glass people to the corresponding association for the others.

2B.13. Indicate on the height-weight plot which points correspond to people with a GPA of greater than or equal to 3.0 and which points to people with a GPA below 3.0. What do you see? Does a student's height and weight appear to be related to the student's GPA? Compute the correlation coefficient and interpret r^2 for the GPA $\geq$ 3.0 people. Compute the correlation coefficient and interpret r^2 for the GPA < 3.0 people. Contrast the association of height and weight for the GPA $\geq$ 3.0 people to the corresponding association for the GPA < 3.0 people.

2B.6 MINITAB APPENDIX

To obtain a scatter plot in MINITAB, you can use the PLOT command. For example, if the numbers 1, 2, 3, 4, 5, 6 and 7 are in column C1 and the numbers 2, 4, 6, 8, 10, 12, 14 are in column C2, then the command

PLOT the data in column C2 versus the data in column C1

or more simply PLOT C2 C1 will yield a scatter plot with the C2 values on the y-axis and the C1 values on the x-axis.

Assume that points 1, 2, 5 and 6 are associated with men and the remaining points are associated with the women. To notate the sex of the individual on the plot, use the LPLOT command. In particular, set the numbers 13, 13, 23, 23, 13, 13, 23 (type HELP LPLOT to see why 13 and 23 were chosen) in column C3 and type

LPLOT the data in column C2 versus the data in column C1 and notate the points using the codes in column C3

or more simply LPLOT C2 C1 C3 .

To compute the correlation between the data in column C1 and the data in column C2, tell MINITAB to compute the

CORRelation between the data in columns C1 and C2

or more simply CORR C1 C2. MINITAB will assume that each observation in column C1 is paired with the corresponding observation in column C2.

2C

The Shape of Things That Are - Part C

Key Words: Chance Variation; Reference Distribution; Statistical Inference.

> If a man stands with his right foot on a hot stove and his left foot in a freezer, some statisticians would assert that, on average, he is comfortable.
>
> -Anonymous

2C.1 INTRODUCTION

To compliment the arsenal of descriptive techniques discussed in Chapter 2A and 2B, it is important to understand something called statistical inference. **Statistical inference** provides tools for assessing whether two or more numbers, two or more stem-and-leaf plots, two or more lines ... are different. Just because two numbers are different, does not imply that the numbers are *statistically* different. Do you remember how you felt when you were asked to analyze the data from your experiment in Chapter 1? You examined the stem-and-leaf plots for the "e" and "r" measurement errors. You noticed certain differences between the two plots. If the experiment were to be repeated, i.e. everyone was again randomized to the two treatments, it is most certain that the group of people that were now asked to count the e's would not be identical to the old "e" group. (It is assumed that the conditions of the experiment are identical for each repetition of the experiment.) As a consequence, it is most unlikely that in a repetition of the experiment, the stem-and-leaf plot of the measurement errors for either the "e" or the "r" group would be identical to the old stem-and-leaf plots. In this way, you can see that associated with this experiment is a certain amount of *chance variation*. To assess whether the "e" and "r" measurement error stem-and-leaf plots actually differ, it would be necessary to take this variability into account.

In this investigation, you will reexamine the measurement error data with an emphasis on statistical inference. The purpose of this investigation is to help you to develop a feel for the effect of chance variation on the process of inference.

2C.2 THE REFERENCE DISTRIBUTION AND STATISTICAL INFERENCE

Several years ago, my wife and I were in the market for a new car. After deciding upon a particular model, we decided to collect data regarding the cost of this particular model. We visited car dealers and obtained a number of quotes on the price of the vehicle. Very soon into our observational study we found that there was significant variability to the price of the vehicle which we wished to purchase. By shopping around, we were able to construct a *reference distribution* for the prices of the car. Once we had an idea of the distribution of prices, i.e. a reference distribution for the prices, we could determine whether a dealer's price was reasonable or possibly very unreasonable. In fact, at one dealership, we were quoted a price for the car which was out of line. We mentioned to the salesperson that the price seemed somewhat discrepant in relation to what other dealers were asking. His response was that if we could purchase the car for less *somewhere else,* then we should go *somewhere else.* As we were walking out, we examined the slip of paper which listed the various options and their cost and we found that there was an error in the salesperson's arithmetic to the effect of $1000.

Statisticians put a great deal of emphasis on reference distributions. Because statisticians are forced to work in the presence of *chance variation* that they must refer to reference distributions. For a statistician, it is not good enough for two numbers to differ. For the statistician, for the two numbers to be statistically different, they must differ beyond what is expected by chance. It is a statistical reference distribution which allows the statistician to gauge whether the results differ beyond what is expected by chance.

2C.2.1 INFERENCE IN THE PRESENCE OF CHANCE

As an example of inference in the presence of chance variation, recall the experiment that you performed in Chapter 1. Everyone in the class was randomized to one of two groups, a group which enumerated the e's in a text and a second group which enumerated the r's in a text. Associated with each person is a measurement error, i.e. the *enumerated* number of e's or r's minus the *true* number. For instance, for one person the measurement error may be 5, while for another it may be -6 and so on. One question of interest is whether counting e's and counting r's differ with regard to accuracy, i.e. whether the average measurement error differs between the two groups.

By randomizing the members of your class to one of the two groups, chance variation was introduced into the experiment. That is, had the experiment been repeated, the difference in the average measurement error may have changed. Possibly the averages would have differed by a half letter or by one letter or by five letters. Rather than simply examine whether the average measurement error differs for the two groups, one must determine whether the averages differ beyond what is expected by chance variation. For the statistician to conclude that the averages differ, the difference in the average measurement error has to be *more extreme than what would be expected by chance.*

2C.2.2 THE STATISTICAL APPROACH

To illustrate how we will address whether the average measurement error differs for the two groups, assume that the observed difference in the average measurement error between the "r" and the "e" group is 1.5 letters. (The number 1.5 was selected for illustration.) How would a statistician assess whether this 1.5 represents a *statistically significant* difference? A statistician would ask the following question:

> Assuming for the moment that there is no difference in the average measurement error between the "r" and "e" groups, what is the probability of observing a difference in the average measurement error of 1.5 letters or more?

To understand how one can determine whether the average measurement error for the two groups differ, one has to understand the three-step process of statistical inference. First, one states a hypothesis about the system which one wishes to study. In particular, the statistical approach (devil's advocate approach) is to hypothesize that there is no difference between the "e" and the "r" group. Next, proceeding under this hypothesis, one constructs the reference distribution and then performs a statistical calculation using the reference distribution. The goal of this calculation is to compute the probability of observing a difference in the average measurement error between the two groups *as extreme or more extreme* than what was actually observed. If the probability of observing a difference of 1.5 letters or more is very *small,* then the (observed) data indicate that there is evidence to *doubt* the assumption that there is no difference in the average measurement error between the two groups.

To elaborate a bit on this last remark, let's consider the situation in which, before you performed the experiment, you were asked to assess the probability of observing a difference between the "r" and the "e" group of 1.5 letters or more. Let us say that on the basis of your hypothesis that there is no difference between the two groups, you conclude that the probability of observing a difference of 1.5 or more letters, *due to chance,* is zero. In other words, what your calculation is telling you is that even given chance, it would be impossible to observe a difference between the two groups as extreme or more extreme than 1.5 letters. Now what if you performed the experiment and found a difference of 4.2 letters? Clearly, you would be forced to conclude that your initial hypothesis upon which your computation is based *is wrong*. In other words, you would conclude that the assumption of no difference between the "e" and the "r" groups, with regard to accuracy, is wrong.

2C.2.3 BUILDING A REFERENCE DISTRIBUTION AND COMPUTING THE PROBABILITY OF BEING EXTREME

In this chapter's investigation, you will be asked to compare the average measurement error for the "e" group to the average measurement error for the "r" group. To assess the statistical significance of this difference, you will empirically build a reference distribution by considering the ways in which the data could have looked. You will then compare the *observed* difference between the average measurement error for the "e" and "r" groups to the *could-have-been-observed (simulated)* differences in an attempt to assess whether the observed difference is unlikely. The probability of interest will be the proportion of could-have-been-observed (simulated) differences which are as extreme or more extreme than the observed difference.

To build a reference distribution for the counting experiment, one must understand four points, the key observations being the clever way one uses the assumption of no difference and the fact that the students were *randomized* to count e's or to count r's.

First, associated with each measurement error is a label: "e" or "r". Here is an example of a class with *four* students, where two students were randomized by the flip of a coin to count r's, two students were randomized by the flip of a coin to count e's, and their associated measurement errors are given as:

Label:	"r"	"e"	"r"	"e"
Measurement Error:	7	1	2	5

One student who counted r's was off by 7 letters, while another student who counted e's was off by 1 letter, and so on. The *observed* difference in the average measurement error (the "r" group average measurement error less the "e" group average measurement error) is 1.5.

Second, if we assume that in counting e's and r's there is no difference with regard to accuracy, then "e" and "r" are only labels which have no effect on the measurement error. That is, under this assumption, the first measurement error would have been 7 whether the student enumerated the e's or the r's, the second measurement error would have been 1 whether the student enumerated the e's or the r's and so on. In other words, by making the assumption that in counting e's and r's there is no difference with respect to accuracy, this implies that the first measurement error was in a sense *predetermined* to be 7, irrespective of whether the student counted e's or r's.

Third, note that by randomizing everyone in the class, only *one* possible outcome of the "e" and the "r" labels was specified, i.e. (r, e, r, e). However, there are other outcomes of the labels *which could have occurred,* assuming that counting e's and counting r's do not differ with regard to accuracy. For example:

Label:	"e"	"r"	"e"	"r"
Measurement Error:	7	1	2	5

is a possible outcome. That is, in this potential outcome, the student who was off by 7 letters would have been randomized by the flip of a coin to count e's, while the student who had a measurement error of 1 would have been randomized to count r's and so on. Remember that under our assumption that "e" and "r" are mere labels which do not differ in their effect on the measurement error, the measurement error was determined *before the experiment was performed.* In other words, our basic assumption is that *a student's error would have been the same whether e's or r's were counted.* Similarly, the outcome

Label:	"r"	"e"	"e"	"r"
Measurement Error:	7	1	2	5

could have been obtained by randomizing the students to count either e's or r's on the basis of a flip of a coin.

Finally, to actually construct the reference distribution, we would perform the following task: For each of the possible outcomes of the "e" labels and the "r" labels, we compute the difference in the average measurement errors ($\bar{x}_r - \bar{x}_e$). For example, in the first alternative outcome (e, r, e, r), the difference in average measurement errors is 3.0 - 4.5 = -1.5 ($\bar{x}_r - \bar{x}_e$). In the second alternative outcome (r,e,e,r), the difference is 6.0 - 1.5 = 4.5. In fact, there are sixteen possible outcomes:

Outcome	*7*	*1*	*2*	*5*	$\bar{x}_r - \bar{x}_e$
1	e	e	e	e	-
2	r	e	e	e	+4.33
3	e	r	e	e	-3.66
4	r	r	e	e	+0.50
5	e	e	r	e	-2.33
6	r	e	r	e	+1.50
7	e	r	r	e	-4.50
8	r	r	r	e	-1.66
9	e	e	e	r	+1.66
10	r	e	e	r	+4.50
11	e	r	e	r	-1.50
12	r	r	e	r	+2.33
13	e	e	r	r	-0.50
14	r	e	r	r	+3.66
15	e	r	r	r	-4.33
16	r	r	r	r	-

(Of these sixteen outcomes, we eliminate the outcomes (e, e, e, e) and (r, r, r, r) because they do not allow for the computation of $\bar{x}_r - \bar{x}_e$.) In this way, if one were to compute the difference in average measurement errors for each of the fourteen *allowable* outcomes and then draw the histogram for these differences, one would obtain the required reference distribution. One would then use this reference distribution to assess whether or not the observed difference of 1.5 is extreme. If relatively few simulated differences are larger than or equal to the observed difference, then either the assumption that the e's and the r's are mere labels is wrong, or an unusually large difference given what would have expected from chance variation was observed. In such a circumstance, one would tend to believe that the assumption of no difference is wrong. In fact, for six out of the fourteen *allowable* outcomes listed in the above table, the associated $\bar{x}_r - \bar{x}_e$ is larger than or equal to 1.5. Thus, in this particular case, there is little evidence against the assumption of no difference.

To summarize the discussion of this section, we wish to determine whether the observed difference between two treatments is more extreme than what we would expect by chance. To accomplish this task, we compare the *observed* results of the experiment to the results of all the experiments *which could have occurred as a result of the randomization, under the assumption of no treatment difference.* If relatively few of these potential outcomes are as extreme or more extreme than what was observed, then either the assumption of no treatment difference is wrong or we observed an unusually large difference given chance. In such an instance, we would tend to believe the former explanation that our assumption of no treatment difference is wrong.

In this chapter's investigation, you will not examine all possible outcomes of the "e" and "r" labels (for a class of 40 students that would involve 2^{40} - 2 enumerations!). Rather you will randomly select a much smaller number of such outcomes and then use these outcomes to *approximate* the reference distribution to determine whether the difference you observed between the "e" and the "r" groups is statistically significant.

The books by Box, Hunter and Hunter (1978, page 95), Lehmann (1975, pages 43 and 106), Edgington (1980) and Cox and Hinkley (1974, pages 196-198) discuss the construction of the reference distribution.

2C.3 MATERIALS

One coin (preferably new and of the same denomination) for each student and the results of the investigation in Chapter 1.

2C.4 EXPERIMENTAL PROTOCOL

This investigation makes use of the data which you generated in Chapter 1. Your instructor will provide you with the results of the investigation in Chapter 1 so that you will be able to perform the following activity:

Activity:

> For each recorded measurement error, flip the coin. If the flip results in a **HEAD,** then assign that measurement error to group "e". (Write the value in the "e" column on the following page.) If the flip results in a **TAIL,** then assign that measurement error to group "r". (Write the value in the "r" column on the following page.)

"e"	"r"

Now compute $\overline{x}_r$ - $\overline{x}_e$ __________________.

Repeat the Activity four more times.

Second Pass

"e"	"r"

Now compute $\overline{x}_r$ - $\overline{x}_e$ ________________.

Third Pass

"e"	"r"

Now compute $\overline{x}_r - \overline{x}_e$ ______________.

Fourth Pass

"e"	"r"

Now compute $\overline{x}_r$ - $\overline{x}_e$ __________________.

Fifth Pass

NAME________________________

"e"	"r"

Now compute $\overline{x}_r - \overline{x}_e$ ________________.

DATA SHEET

Record the five simulated differences below. Detach this page and hand in your results.

__________ __________ __________ __________ __________.

2C.5 ANALYSIS

Your instructor will provide you with the simulated differences which your classmates generated, so that you will be able to answer the following questions: Questions 2C.1-2C.4 address the statistical significance of the observed difference in the "e" and the "r" measurement errors.

2C.1. What was the *observed* difference in the average measurement error between the "e" and the "r" group? ($\bar{x}_r$ - $\bar{x}_e$)

2C.2. Draw a histogram for the simulated differences. (This is the reference distribution to be used to assess whether there is a difference in the average measurement error between the "e" and the "r" groups.) What do you see? Describe the histogram with regard to SLAS.

2C.3. What percentage of the simulated differences are greater than or equal to the observed difference? What is your conclusion regarding the difference in the average measurement error between the "e" and the "r" groups?

2C.4. If you conclude that the average measurement error differs between the two groups, can you ascribe this difference entirely to the phonetic nature of the letters? In other words, can you say that the average measurement error differs between the two groups because the letter "r" has a distinctive sound, while the letter "e" is a somewhat "softer" letter? Is it possible that the measurement error also depends on the *frequency* of the occurrence of the letter? How would you design an experiment to unravel the effect of the phonetic aspects of the letter and the effect of the frequency of the occurrence of the letter?

2C.6 MINITAB APPENDIX

Given the observed measuremnet errors in column C1, MINITAB can be used to randomize the numbers to the "e" or to the "r" group and then compute $\bar{x}_r$ - $\bar{x}_e$. In particular, if there are 41 numbers in C1, then the command sequence:

```
RAND 41 C2;
BERN 0.5.
COPY C1 C3;
USE C2=1.
COPY C1 C4;
USE C2=0.
```

will randomly assign errors to column C3 ("r" column) and to column C4 ("e" column). The command sequence

```
AVER C3
AVER C4
```

will then compute $\overline{x}_r$ and $\overline{x}_e$.

2C.7 REFERENCES

1. Box, G. E. P., Hunter, W. G. and Hunter, J. S. *Statistics for Experimenters,* John Wiley and Sons, New York, 1978.

2. Cox, D. R. and Hinkley, D. V. *Theoretical Statistics,* Chapman and Hall, London, 1974.

3. Edgington, E. S. *Randomization Tests,* Marcel Dekker, Inc., New York, 1980.

4. Lehmann, E. L. *Nonparametrics,* Holden-Day, Inc., San Francisco, 1975.

3A

Are You Well Calibrated?

Key Words: Objective/Subjective Probability.

If you think there is safety in numbers, try playing roulette.
-Anonymous

3A.1 INTRODUCTION

What does it mean that the probability of the statement "The tossed coin will turn up heads" is one-half? Either the coin will turn up heads in which case the statement is true or it will turn up tails in which case the statement is false. Why should your position in time and space effect the truth of the proposition? Even more perplexing is the argument that coin tossing is a random experiment. There is nothing random about tossing a coin. It is the initial forces on the coin, the nature of the surface upon which the coin lands and Newton's laws which determine how the coin will turn up. There is nothing random about Newton's laws - after all, applications of these laws took astronauts to the moon and took unmanned probes to the edge of the solar system. After these few comments, you can begin to see that *probability* and *randomness* are not trivial notions but require careful consideration.

3A.1.1 THE NOTION OF PROBABILITY

A number of philosophers have attempted to define the term *probability*. de Finetti (1978) surveys two general approaches to probability: the *subjective* approach and the *objective* approach. Regarding the subjective approach, de Finetti begins with the simple question of "What can be said of an event?" de Finetti notes that for a particular individual, at a given moment, "an event may be certain, impossible, or dubious". That is, before the coin is tossed, either the coin will land heads up or tails up, but one cannot say with certainty which result will occur before the experiment is performed. If one were asked, before the coin is tossed, how the experiment would turn out, one could not respond that the event "The tossed coin will turn up heads" is impossible nor certain to occur. Rather, based on an assessment of one's beliefs, one would develop some degree of confidence regarding the event "The tossed coin will turn up heads". This degree of confidence is the subjective probability. To de Finetti, this void between the certain and the impossible, "the dubious, is split into a continuous range of degrees of doubt, or degrees of belief, which are subjective probabilities." **Subjective** probability represents the degree of an individual's belief in the outcome of an event.

In the objective approach, probabilities are notions, like mass or distance, which are inherent in nature. As B. V. Gnedenko (1962) points out: *"probabilistic judgments express certain objective properties of the phenomena that are studied"*. In the objective approach, it is the act of tossing the coin, not the observer who observes the experiment, which is imbued with probability.

Gnedenko begins with the observation that there are many phenomena for which, *whenever a particular set of conditions is realized,* an event may or may not occur. In other words, for a given set of conditions, one cannot predict the behavior of certain phenomena with certainty. For example, whenever one tosses a coin, sometimes it will land heads up and sometimes not and one cannot predict which will occur. **Objective** probability is a quantitative index of the particular phenomenon under study. If one were to repeatedly toss a coin under a fixed set of conditions, the ratio of the frequency of observing a head to the total number of tosses would tend to approach some value. To Gnedenko, this value is an index of the coin toss experiment and this value represents the probability of observing a head.

You will return to this question of determining a probability by the repetition of an experiment under a fixed set of conditions in Chapter 4, when you perform a classical coin tossing experiment. However, before proceeding, it is noted that, as with the subjective approach, there are many ways of approaching objective probability. von Mises (1957), as well as Bernstein (1946), discuss alternative views and their work is reviewed in Maistrov (1974).

Having looked at two general approaches to defining probability, how can one understand *randomness?* To the subjectivists, randomness is a function of one's doubts about the world. Because a person may have doubts about the truth of the event "The tossed coin will come up heads", one might associate the term "random" with this event. However, if it is believed that this event is certain, then tossing the coin would not be "random".

To the objectivists, randomness is a function of the degree to which one can specify the conditions under which one observes an event. In particular, according to Gnedenko (1962): "An event that may or may not occur when the set of conditions C is realized, is called random." Under the (rather vague) set of conditions that I place a coin on my thumb nail and then flip the coin into the air, the coin may land heads up or it may land tails up (excluding the situation where it rolls away!). Under such a set of conditions, it is agreed that we are observing a random phenomenon, in that the result of heads or tails cannot be predicted. However, if one restricts the conditions under which the experiment is performed, then the phenomenon may no longer be random. In particular, if a machine was developed which could deliver a very well defined force to a given point on the coin, then conceivably one could always observe heads. In such a case, one would be hard-pressed to prove that such an experiment results in random occurrences. However, such a coin tossing machine would have to be very precise, for in coin tossing, even very small changes in the way the force is delivered to the coin would completely destroy the predictability of the phenomenon. It is in this way that coin tossing results in random outcomes. Coin tossing is "random" for very small changes in the initial conditions of the system have dramatic effects on the outcome of the toss.

At this point, there is no universally accepted approach to probability - each theory has its strengths and weaknesses. (See de Finetti (1978) and Maistrov (1974) for a critique of the objective approach and von Mises (1957) for a critique of the subjective approach.) For example, subjective probability may be useful in describing an unrepeatable process. However, since it is personal, your subjective probability may or may not be congruent with my assessment nor with the long term relative frequency of the event should this frequency exist. In this regard, a person is called **calibrated** if, over the long-run, for all events assigned a particular subjective probability, the percentage of events which occur equals the assigned subjective probability value (Lichtenstein, Fischhoff and Phillips, 1983). The goal of this chapter's investigation will be to examine how well your classmates' subjective assessment of the probability of various events matches the observed relative frequency of these events.

3A.2 THE BASICS - COMPUTING PROBABILITIES

In a survey in which students were asked to record their sex and report whether or not they considered themselves to be a smoker, 43 responses were obtained. The data are summarized in the following $2x2$ frequency table.

	Non-Smokers	Smokers
Men	24	3
Women	12	4

As shown in the table, there were four types of people: men smokers, men non-smokers, women smokers and women non-smokers. As also seen in the table, of the 43 respondents, 24 men claimed to be non-smokers, 3 men claimed to be smokers, 12 women claimed to be non-smokers and 4 women claimed to be smokers.

Now suppose that the sex and smoking status of each of the 43 respondents in the survey were written on a ticket and put into a hat. Then the tickets were mixed and one was pulled out of the hat. What is the probability that the pulled ticket said "Woman, Non-Smoker"? In other words, what is the probability of the event

"The selected ticket is labeled 'Woman, Non-Smoker'"?

(The word **event** is used to mean a statement about the world.) An objective probabilist would argue as follows: If each ticket is equally likely to be drawn, and one were to repeatedly draw tickets from the hat (returning each ticket to the hat before the next ticket is drawn), then *in the long-run each ticket would be drawn about once in forty-three draws*. In other words, each ticket will be selected with a probability of $\frac{1}{43}$. Now notice that out of the 43 tickets in the hat, there are 12 tickets which are labeled "Woman,

Non-Smoker", so that in the long-run one can expect to select such a ticket twelve times in forty-three draws. Thus, the objectivist would answer that the probability of the event of interest is $\frac{12}{43}$ or 0.28.

A rational subjectivist might also come to 0.28 if this person is convinced that there is a *symmetry* present in the experiment, i.e. that each ticket is equally likely to be drawn. In the situation where one believes that all tickets are equally likely to be drawn, one is forced to assign equal "degrees of belief" or probability to each of the tickets. In this way, because there are 43 tickets, to the "degree of belief" that any particular ticket will be drawn, the rational subjectivist would assign the value $\frac{1}{43}$. Since there are 12 tickets labeled "Woman, Non-Smoker", to the "degree of belief" that any ticket labeled "Woman, Non-Smoker" will be drawn, the rational subjectivist would assign the value 12 x $\frac{1}{43}$ or $\frac{12}{43}$. Note that both the subjectivist and objectivist arrive at the figure 0.28. For the subjectivist, this figure quantifies a *belief,* while for the objectivist this figure is an *index* of process which is to be repeated under identical conditions.

It turns out that people like to abbreviate the statement "The probability of the event A" with some sort of notation. Some people use p(A), some use P(A), while some use Pr(A). We will use the notation p(A) to denote the probability of the event A. Thus, using the data from the above survey, if one lets A denote the event "The selected ticket is labeled 'Woman, Non-Smoker'", then p(A) = 0.28. Probability is discussed in more detail in Freedman, Pisani and Purves (1978).

3A.3 MATERIALS

Each student should have a copy of the forty trivia questions.

3A.4 EXPERIMENTAL PROTOCOL NAME______________________

In this investigation, you will examine how well your assessment of a subjective probability matches the corresponding empirically observed proportion. For each of forty statements you will be asked to respond true or false, *as well as record your subjective probability that the response is correct.* **Please limit your numeric response for each subjective probability to a value in the range 50% - 100%.** If you are *absolutely sure* of your response, then you would record 100%, while a *guess* would be 50% since either response (TRUE or FALSE) is equally likely.

1. West Germany is the country to the south of Denmark.

TRUE_____ FALSE_____ p_____

2. According to Karl Marx, the opium of the people is money.

TRUE_____ FALSE_____ p_____

3. "Have you never been in love?" is the title of a song sung by Olivia Newton-John.

TRUE_____ FALSE_____ p_____

4. The amount of water in the human body averages 35%.

TRUE_____ FALSE_____ p_____

5. A horse named Octavian won the Kentucky Derby five times.

TRUE_____ FALSE_____ p_____

6. Julie Andrews won the award for Best Actress for her role in "Mary Poppins".

TRUE_____ FALSE_____ p_____

7. Carlo Collodi is the author of "The Adventures of Pinocchio".

TRUE_____ FALSE_____ p_____

8. A B25 bomber once crashed into the Empire State Building.

TRUE_____ FALSE_____ p_____

9. Antibiotics work by killing viruses.

TRUE_____ FALSE_____ p_____

10. Mickey Rooney appeared in the film "The Black Stallion".

TRUE_____ FALSE_____ p_____

11. Dustin Hoffman appeared in the film "Lenny".

TRUE_____ FALSE_____ p_____

12. The earliest human beings lived at the same time as the dinosaurs.

TRUE_____ FALSE_____ p_____

13. The Panama canal is longer than the Suez Canal.

TRUE_____ FALSE_____ p_____

14. "Little Boy" is the code name for the bomb dropped on Hiroshima.

TRUE_____ FALSE_____ p_____

15. Nicholaus Copernicus was Greek.

TRUE_____ FALSE_____ p_____

16. Willy Lomack is a character in the play "Death of a Salesman"

TRUE_____ FALSE_____ p_____

17. The oxygen we breathe comes from plants.

TRUE_____ FALSE_____ p_____

18. Glen Campbell won a Grammy Award for the song "By the time I get to Phoenix".

TRUE_____ FALSE_____ p_____

19. Diana Ross starred in the movie "Lady Sings the Blues".

TRUE_____ FALSE_____ p_____

20. Vita Blue won the American League Cy Young Award.

TRUE_____ FALSE_____ p_____

NAME________________________

21. Karl Benz built the first car with an internal combustion engine.

TRUE_____ FALSE_____ p_____

22. Redd Foxx's original name was John Sanford.

TRUE_____ FALSE_____ p_____

23. Alexander the Great defeated Darius in 333 B.C.

TRUE_____ FALSE_____ p_____

24. Art Linkletter's picture is on the $50 bill in the game of Password.

TRUE_____ FALSE_____ p_____

25. According to Greek mythology, Pandora was the first woman on earth.

TRUE_____ FALSE_____ p_____

26. "Are You Lonesome Today?" is the title of a song sung by Elvis Presley.

TRUE_____ FALSE_____ p_____

27. The cheetah can run faster than 65 miles per hour.

TRUE_____ FALSE_____ p_____

28. French-fried potatoes were invented by the Belgians.

TRUE_____ FALSE_____ p_____

29. An ant can pick up a stone in excess of twenty times its own weight.

TRUE_____ FALSE_____ p_____

30. Minsk is the oldest city in Russia.

TRUE_____ FALSE_____ p_____

31. Mount Holyoke was the first women's college in the United States.

TRUE_____ FALSE_____ p_____

32. The thermometer was invented in the eighteenth century.

TRUE_____ FALSE_____ p_____

33. The Beatles' first appearance in the United States was on the Ed Sullivan show.

TRUE_____ FALSE_____ p_____

34. Lou Gehrig was one of the first five players elected to the Baseball Hall of Fame.

TRUE_____ FALSE_____ p_____

35. Godzilla appeared on the covers of both *Time* and *Newsweek.*

TRUE_____ FALSE_____ p_____

36. Hans Rey is the author of "Curious George".

TRUE_____ FALSE_____ p_____

37. Joe Frazier was the heavyweight boxing champion from 1973-1974.

TRUE_____ FALSE_____ p_____

38. South Carolina was the first state to secede from the Union.

TRUE_____ FALSE_____ p_____

39. A star's twinkle is due to the earth's atmosphere.

TRUE_____ FALSE_____ p_____

40. Becquerel shared the Nobel Prize for the discovery of radioactivity.

TRUE_____ FALSE_____ p_____

Record your sex. __________ **MALE** __________ **FEMALE**

Record your GPA. __________

Hand in pages 57-60, so that the class results may be pooled.

3A.5 ANALYSIS

Your instructor will provide you with the class results, so that you may answer the following questions. In particular, you will be given a series of seven $5x3$ frequency tables. The first row of the table will correspond to those responses assigned a subjective probability in the interval [0.5,0.6). The second row corresponds to those responses assigned a subjective probability in the interval [0.6,0.7); the third [0.7,0.8); the fourth [0.8,0.9); and the fifth [0.9,1.0). The first number in each row is the average of the subjective probabilities for the particular range (e.g. the first number in the first row is the average of the subjective probabilities in the range [0.5,0.6)), while the second number in that row is the median of the subjective probabilities (e.g. the second number in the first row is the median of the subjective probabilities in the range [0.5,0.6)). The third number in each row is the observed rate of correct responses for the assessments of that row.

The first table will be pooled across the results of all the members of your class, the second table will use the women's results, the third table will use the men's results, the fourth table will use the results of people whose GPA > 3.25, the fifth table will use the GPA ≤ 3.25 results, the sixth table will use the results of the "easy" questions (i.e. those questions where seventy-five percent or more of the responses are correct), while the final table will use the results of the "hard" questions (i.e. those questions where less than seventy-five percent of the responses are correct).

3A.1 For the class results, plot the observed rate of correct response versus the average subjective probability. Draw the 45° line on the plot. What does this line represent? Describe what you see.

3A.2 Repeat 3A.1 for the women's data.

3A.3 Repeat 3A.1 for the men's data.

3A.4 Repeat 3A.1 for the GPA>3.25 data.

3A.5 Repeat 3A.1 for the GPA≤3.25 data.

3A.6 Repeat 3A.1 for the "hard" questions.

3A.7 Repeat 3A.1 for the "easy" questions.

3A.8 Based on your responses to 3A.1-3A.7, do the subjective probability assessments of your classmates tend to match the empirical frequencies? In other words, is your class *calibrated?* Are certain subsets of your class better calibrated than other subsets? Is the degree of calibration related to the difficulty of the question?

3A.6 REFERENCES

1. Bernstein, S. N. *Teoriya Veroyatnostei,* Moscow-Leningrad, 1946.

2. de Finetti, B. "Probability: Interpretations", in *International Encyclopedia of the Social Sciences,* vol. 12, pp. 496-504, Macmillan, New York, 1978.

3. Freedman, D., Pisani, R. and Purves, R. *Statistics* Norton, New York, 1978.

4. Gnedenko, B. V. *The Theory of Probability,* Chelsea, New York, 1962.

5. Lichtenstein, S., Fischhoff, B. and Phillips, L. D. "Calibration of Probabilities: The State of thc Art to 1980", in *Judgment Under Uncertainty: Heuristics and Biases,* Cambridge University Press, Cambridge, 1983.

6. Maistrov, L. E. *Probability Theory: a Historical Perspective,* Academic Press, New York, 1974.

7. von Mises, R. *Probability, Statistics and Truth,* George Allen and Unwin Ltd, London, 1957.

3B

The Vulcan Salute

Key Words: Sensitivity / Specificity; False Positive and False Negative Probabilities; Measures of Association; Odds Ratio; Correlation Coefficient.

> Life is a process of acquiring more and more numbers until one's number is up.
>
> -Anonymous

3B.1 INTRODUCTION

Modern civilization has witnessed the development of a number of diagnostic tests. In general, these diagnostic tests are used for determining whether a given subject or unit has a dichotomous attribute before the attribute appears or in situations where the attribute is difficult to detect. For example, to judge whether a person is guilty or innocent, lying or telling the truth, the polygraph test was developed. Modern medicine has given us the Pap smear as an early detection device for the presence or absence of cancer of the cervix, while the chest X-ray has been used for early diagnosis of tuberculosis. The 10 o'clock weather reporter relies on a range of gadgets for predicting whether or not it will rain the next day.

For such tests to be useful diagnostic tools, they must be valid. To be valid, the test should have a low *false negative probability* - that is, the test should avoid labeling a subject or unit with the attribute as not having the attribute. For example, one would want a test for cancer which would avoid labeling a person who actually has cancer as being healthy. Similarly to be valid, a test must have a low *false positive probability* - that is, the test should avoid labeling a subject or unit without the attribute as having the attribute. For example, a polygraph test should avoid labeling an innocent person as guilty. The goal of this chapter's investigation is to examine the validity of a test for right or left handedness.

3B.2 NUMERICAL SUMMARIES OF 2x2 TABLES

To analyze the data from this investigation, several terms are defined.

The **false positive probability** of a test is the probability of an incorrect decision when the subject being classified DOES NOT have the attribute.

The **false negative probability** of a test is the probability of an incorrect decision when the subject being classified DOES have the attribute.

The **sensitivity** of a test is the probability of a correct decision by the test given that the subject or unit DOES have the attribute.

The **specificity** of a test is the probability of a correct decision by the test given that the subject or unit DOES NOT have the attribute.

In terms of the generic $2x2$ frequency table,

Prediction	Attribute Present	Attribute Absent
Present	a	b
Absent	c	d

the false positive probability is given by $\frac{b}{b+d}$, the false negative probability is given by $\frac{c}{c+a}$, the sensitivity is given by $\frac{a}{a+c}$, while the specificity is given by $\frac{d}{b+d}$.

Two measures of association for the $2x2$ table are now defined.

Given the generic $2x2$ frequency table, the **odds ratio** is $\frac{a/b}{c/d} = \frac{ad}{bc}$.

An odds ratio close to 1.0 indicates nil association, an odds ratio close to 0.0 indicates a strong *negative association,* while a large odds ratio indicates a strong *positive association.* In this regard, note that when b or c is equal to zero, the odds ratio is undefined.

The **correlation coefficient** for a $2x2$ table is

$$r = \frac{ad - bc}{\sqrt{(a+b)(d+b)(a+c)(d+c)}} \ .$$

The correlation coefficient takes values between -1 and +1. A value of -1 indicates strong *negative association,* a value of 0.0 indicates no association, while a value of +1 indicates a strong *positive association.* The number r^2 quantifies the proportion of variability in one variable which is explained by knowledge of the second variable and vice versa.

These terms and measures of association are discussed in further detail in Brown and Hollander (1977).

3B.3 MATERIALS

Each student should have two rubber bands, a pen or pencil and the "hand sheets".

3B.4 EXPERIMENTAL PROTOCOL

In this experiment, you will assess the usefulness of a test for left- or right-handedness.

1. Spread apart the fingers of your RIGHT hand so that a V is formed with the ring and little finger in one branch of the V and the remaining fingers (middle finger, forefinger and thumb) in the second branch. Make sure that the fingers in each branch of the V remain parallel and together.

2. Place your RIGHT hand (with the fingers in the V formation) onto the RIGHT HAND SHEET and trace the outline of the V onto the sheet of paper. The palm of the hand and both branches of the V should touch the paper. You should NOT stretch the fingers of your right hand with your left hand, rather the fingers should be stretched naturally. If you are having difficulty spreading your fingers, then you may wish to tie the ring and little finger together with a rubber band and the remaining fingers together using the other rubber band.

3. Now spread apart the fingers of your LEFT hand so that a V is formed with the ring and little finger in one branch of the V and the remaining fingers (middle finger, forefinger and thumb) in the second branch. Make sure that the fingers in each branch of the V remain parallel and together.

4. Place your LEFT hand (with the fingers in the V formation) onto the LEFT HAND SHEET and trace the outline of the V onto the sheet of paper. The palm of the hand and both branches of the V should touch the paper. You should NOT stretch the fingers of your left hand with your right hand, rather the fingers should be stretched naturally. If you are having difficulty spreading your fingers, then you may wish to tie the ring and little finger together and the remaining fingers together using the rubber bands.

5. Now compare the RIGHT and LEFT HAND SHEETS to determine which V has the **wider angle.** Record your result on the data sheet.

RIGHT HAND SHEET

Remove sheet from book before tracing.

Place the base of your right palm on this line and trace the V onto this page.

LEFT HAND SHEET

Remove sheet from book before tracing.

Place the base of your left palm on this line and trace the V onto this page.

DATA SHEET **NAME**____________________

1. Record the result of the comparison of the V's.

_________ **RIGHT V BIGGER** _________ **LEFT V BIGGER** _________ **SAME**

2. Record your handedness. (Which hand do you use to write?)

_________ **RIGHT-HANDED** _________ **LEFT-HANDED**

3. Record whether you used the rubber bands.

_________ **USED RUBBER BANDS** _________ **DID NOT USE RUBBER BANDS**

4. Record your sex.

_________ **MALE** _________ **FEMALE**

5. Do you play a musical instrument which has a keyboard (such as a piano or organ) or do you type on a regular basis?

_________ **YES** _________ **NO**

6. In your opinion, is *RIGHT V BIGGER* a predictor of *right-handedness?*

_________ **YES** _________ **NO**

Hand in this page, so that the class results may be pooled.

3B.5 ANALYSIS

Your instructor will provide you with the class results, so that you may answer the following questions. In particular, you will be given a series of nine $2x2$ frequency tables. All tables in the series will be of the form,

	TRUE RIGHT-HANDED	TRUE LEFT-HANDED
PREDICTED RIGHT-HANDED		
PREDICTED LEFT-HANDED		

where one table will summarize all the participants in the study, one table will correspond to the women, one to the men, one to those who used rubber bands, one to those who did not use rubber bands, one to those who use a keyboard, one table to those who do not use a keyboard on a regular basis, one to those that believe that *right V bigger* is predictive of right-handedness and one table to those who do not believe that *right V bigger* is predictive of right-handedness. Your instructor will indicate how the data are to be used to form a prediction. For the purpose of your analysis of the data, treat right-handed as the attribute, i.e. "left-handed" implies that the attribute is absent while "right-handed" implies that the attribute is present.

The idea of this diagnostic test is due to Mubin Syed, who was seven years old when he discovered this test. According to the February 20, 1974 issue of *Current Science,* Mubin discovered this test while watching the television program *Star Trek.* The V-shaped hand sign was used by one of the characters on the television program, Mr. Spock, to greet his fellow Vulcans. A further discussion of this diagnostic test can be found in Syed (1973).

In questions 3B.1-3B.4 you will be asked to investigate the validity of Mubin's approach to predicting handedness. You will examine the false positive and false negative rates, the sensitivity and specificity, and the association in the class results. You will also examine various subgroups of the data to determine whether Mubin's approach performs better in some subgroups than in others. In questions 3B.5-3B.7, you are asked to consider alternative experimental designs to investigate the validity of Mubin's rule.

3B.1. Compute the false positive and false negative probabilities for all students, for the men, for the women, for those who used rubber bands, for those who did not use rubber bands, for those who use a keyboard, for those who do not use a keyboard, for those who believe *right V bigger* to be a predictor of right-handedness and for those who do not believe *right V bigger* to be a predictor of right-handedness. How well does each group tend to do? Do the men and women appear to differ? Do the keyboard / no keyboard groups differ? Do the believers differ from the non-believers? How about those who did and those who did not use rubber bands?

3B.2. Compute the sensitivity and specificity of the test for all students, for the men, for the women, for those who used rubber bands, for those who did not use rubber bands, for those who believe *right V bigger* to be a predictor of right-handedness, for those who do not believe *right V bigger* to be a predictor of right-handedness, for those who use a keyboard and for those who do not use a keyboard. How well does each group tend to do? Do the men and women appear to differ? Do the keyboard / no keyboard groups differ? Do the believers differ from the non-believers? How about those who did and those who did not use rubber bands?

3B.3. Compute the odds ratio and the correlation coefficient to assess the association between predicted handedness and true handedness for all students, for the men, for the women, for those who used rubber bands, for those who did not use rubber bands, for those who believe *right V bigger* to be a predictor of right-handedness, for those who do not believe *right V bigger* to be a predictor of right-handedness, for those who use a keyboard and for those who do not use a keyboard. How well does each group tend to do? Do the men and women appear to differ? Do the keyboard / no keyboard groups differ? Do the believers differ from the non-believers? How about those who did and those who did not use rubber bands?

3B.4. What would be my false positive probability, false negative probability, sensitivity, specificity, odds ratio and correlation coefficient, if I were to label everyone in your class as a "righty". Would I have done better or worse than Mubin?

3B.5. This experiment involved a somewhat difficult hand maneuver, as well as the comparison of the angle between two sets of possibly very wiggly lines. Is it possible that personal preconceptions as to how the data *should look* affected the individual responses? How can you assess whether, in fact, this distortion occurred? What other factors may have distorted the results of the experiment?

3B.6. In an alternative design, each participant in the study would sit before a rater (who was blind to the participant's handedness) and the rater would make the determination of *right V bigger, left V bigger,* or *same.* Discuss the benefits and drawbacks of this design. Compare this design to the design which was actually implemented.

3B.7. In suggesting a design to evaluate the validity of Mubin's test, do you feel it necessary to randomize the order in which the V's are measured?

3B.8. Based on your responses to questions 3B.1-3B.5, is Mubin's test a *valid* approach for predicting handedness? Does his test appear to work better for some subgroups than other subgroups?

3B.6 REFERENCES

1. Brown, B. and Hollander, M. *Statistics: A Biomedical Introduction,* John Wiley, New York, 1977.

2. Syed, I. B. "A Simple Test for Determining Left-Handedness", *New England Journal of Medicine, 289,* 1205, 1973.

4

Flip Out!

Key Words: Frequentism; Binomial Distribution; Normal Distribution; Central Limit Theorem.

17 is *the* one arbitrary random number. It is also called a "Feller" number. Feller, who was a Princeton Professor, but is now dead, always used to say, "Pick any random number, say 17." The point is, if you're going to pick an arbitrary random number, you don't want it to be even, and you want it to be less than 20. 19 is pushing it a little. Also, it better not be divisible by 5 if it's going to be random. 1 is obviously out. 3 is the Trinity; that's no good. 5 is divisible by 5, so we don't want that. 7 is lucky; 9 is a perfect square. 11 is also lucky, as in "7 come 11." 13 is *unlucky,* and 15 is divisible by 5. We've already thrown out all the evens, so 17 is the only number that's left.

-P. Benacerraf

4.1 INTRODUCTION

The investigation that will be performed in this chapter is a coin tossing experiment. More precisely, you will be asked to flip and to spin a coin. Based on the data which you collect, you will describe your results and then attempt to assess whether flipping a coin leads to the same results as spinning a coin. Following a brief discussion of the Frequentist approach to probability, the binomial distribution will be introduced as a technique for describing the results of a finite coin tossing experiment. The normal distribution and the Central Limit Theorem will then be considered.

4.2 THE FREQUENTIST APPROACH TO PROBABILITY

Consider the situation in which a coin is repeatedly flipped under the *same* conditions. According to the Frequentist interpretation of probability, if one were to consider the proportion of times in n tosses in which the event "The tossed coin lands heads up" occurs, then this proportion would converge to a fixed number p (which shall be referred to as the probability of the event of interest), as the number of tosses (n) increases.

To illustrate how this movement toward "a fixed number p" might look for one coin tossing experiment, assume that one repeatedly tosses a *fair* coin and that one plots the history of this experiment. In particular, assume one tosses a coin 4500 times and then plots the error, i.e. the proportion out of n tosses which land "heads up" minus 1/2, versus the number of tosses (n), as:

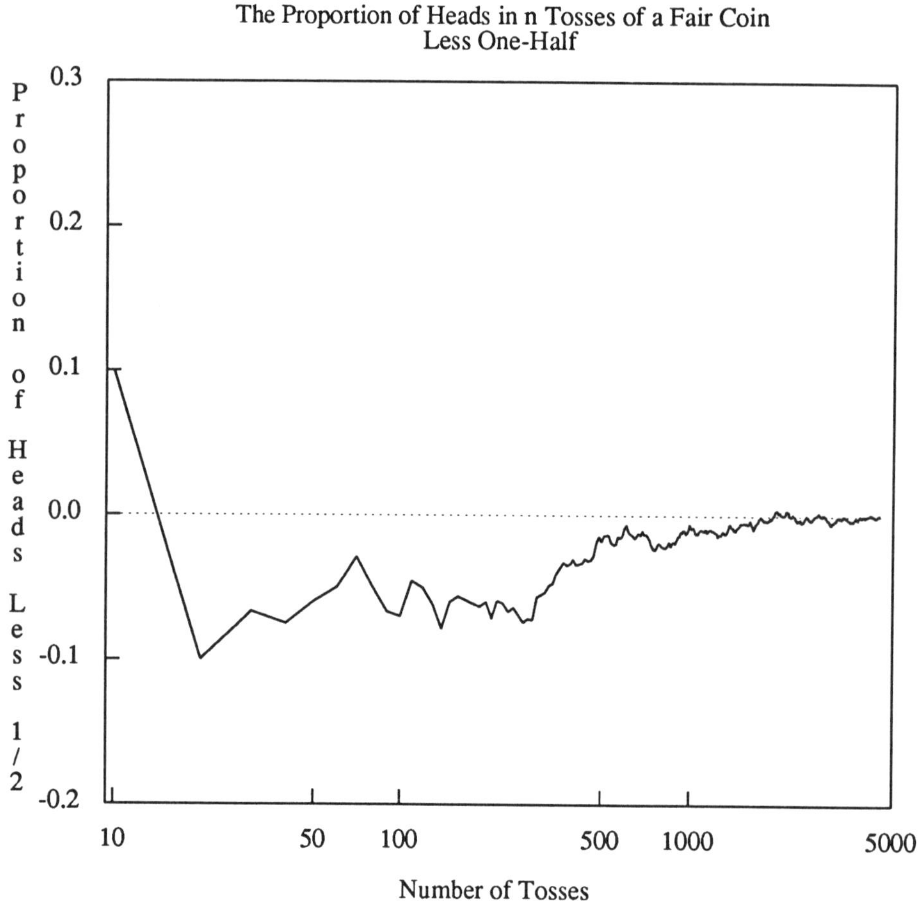

For example, in the first ten flips, six heads were observed so that the observed proportion of heads was 0.6 (6 / 10) and 0.1 (0.6-0.5) versus 10 was plotted. However, in the next ten flips, only two more heads were observed yielding an observed proportion of heads of 0.4 (8 / 20), so -0.1 (0.4-0.5) versus 20 was plotted. Continuing in this way, the entire plot was obtained.

From this plot, one can see that the proportion of tosses resulting in heads fluctuates in value and initially the fluctuations are quite large. However, as the number of tosses of the *fair* coin increases, the probability of a large error tends to approach zero. The proportion of heads in n tosses of a coin converges to the probability of observing a head in the toss of a coin. To the Frequentists, *the accumulation of more data leads to greater precision in the understanding of the behavior of a random process.*

4.3 THE BINOMIAL DISTRIBUTION

The goal of this section is to provide a means for describing the distribution of the outcomes of n tosses of a coin exactly.

To begin, toss a coin five times, where the probability of heads is equal to p and the probability of tails is equal to $q = 1-p$. (If the coin is *fair,* then $p = 1/2 = q$.) One could obtain no heads, one head, two heads, three heads, four heads or five heads. Note that if no heads occur, then one observed

T, T, T, T, T .

What would be the probability of observing such a sequence of no heads? That is, what would be the probability of observing a tail on the first toss *and* a tail on the second toss *and* a tail on the third toss *and* a tail on the fourth toss *and* a tail on the fifth toss? If the tosses are *independent,* then the *multiplication rule* (see Freedman, Pisani and Purves, 1978) indicates that the probability of the event of observing tails on all five tosses (no heads) is given as

$$q \cdot q \cdot q \cdot q \cdot q = q^5 .$$

What is the probability of observing one head? In the situation of one head in five tosses, one could have observed

H, T, T, T, T or

T, H, T, T, T or

T, T, H, T, T or

T, T, T, H, T or

T, T, T, T, H.

Since the probability of observing a head is p and the probability of observing a tail is q, the multiplication rule implies that (one assumes independence of the tosses to apply this rule), the probability of observing any one of these sequences is

$$q \cdot q \cdot q \cdot q \cdot p = q^4 p.$$

Notice that there are five ways in which one head could have been observed in five tosses of a coin and that the occurrence of any one of these five outcomes *precludes* the occurrence of any of the other four outcomes. Thus, the probability of obtaining four tails and one head in a sequence of five tosses of a coin is given by

$$5 \cdot q \cdot q \cdot q \cdot q \cdot p = 5q^4 p.$$

To motivate the binomial distribution as a means for computing the probability of observing k heads in n tosses of a coin, it is helpful to break up this problem into two components. First, in n tosses of a coin, one must know how to compute the probability of observing a particular sequence of k heads and $n-k$ tails. Next, one must know how to calculate how many such sequences there are of k heads and $n-k$ tails.

PART #1- THE PROBABILITY OF A PARTICULAR SEQUENCE

If the tosses of the coin are independent, then by the multiplication rule the probability of observing a *particular* sequence of k heads and $n-k$ tails is

$$p^k q^{(n-k)}.$$

For example, in five tosses of a coin, the probability that the sequence

H, H, T, T, T

will be observed is $p^2 q^3$.

PART #2 - HOW MANY SUCH SEQUENCES ARE THERE?

How many such sequences are there of k heads and $n-k$ tails in n tosses of a coin? To compute this number, use the *binomial formula* which indicates that there are

$$\binom{n}{k} = \frac{n!}{(n-k)!\,k!}$$

such sequences, where $n! = n(n-1)(n-2)\cdots 1$, $1! = 1$, $0! = 1$ and one reads

$$\binom{n}{k}$$

as "n choose k". (For example, 5! = 5x4x3x2x1 = 120 and 4! = 4x3x2x1 = 24.) Rather than enumerate the distinct sequences of one head in five tosses of coin, one can use the binomial formula as a short cut to indicate that there are "5 choose 1" or

$$\binom{5}{1} = \frac{5!}{(5-1)!\,1!} = \frac{5x\,4x\,3x\,2x\,1}{(4x\,3x\,2x\,1)(1)} = 5$$

such sequences.

Parts #1 and #2 are combined to give:

The probability that one will observe k heads in a sequence of n independent tosses is given by the **binomial distribution:**

$$\binom{n}{k} p^k q^{(n-k)} \text{ , } k = 0, 1, 2, \ldots, n,$$

where the probability of heads at *any toss is equal to p.*

For example, in tossing a *fair* coin five times, the probability of two heads is equal to

$$\binom{5}{2} 0.5^2 0.5^3 = 10 \cdot (\frac{1}{2})^5 = 0.3125,$$

that is "5 choose 2" times the probability of the sequence HHTTT when tossing a *fair* coin.

4.4 THE NORMAL DISTRIBUTION

In this section, we will present the normal distribution and show how to work with the normal table (i.e. compute areas under the the normal distribution).

4.4.1 THE NORMAL DISTRIBUTION AND NORMAL TABLE

The **normal distribution** given as:

$$y = \frac{1}{\sqrt{2\pi}} e^{-\frac{1}{2}x^2} \quad \text{for } -\infty < x < \infty,$$

where π is a number which is approximately 3.14159, e is a number whose first six digits are 2.71828 and $\sqrt{2}$ is about 1.41421 . The graph of the normal distribution as a function of x (over the range [-4.0,4.0]) is given as:

The Graph of the Normal Distribution

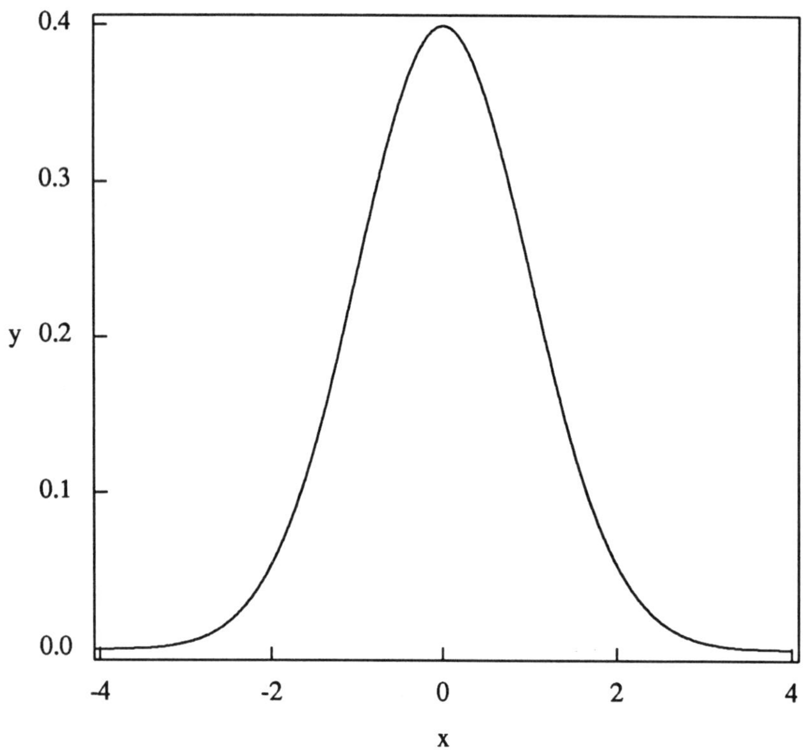

The normal distribution is unimodal and centered at zero. In fact, one very important property of the normal distribution is that it is symmetric about zero. It is also pointed out that the area under the entire distribution (from $-\infty$ to ∞) is equal to 1.0, while the area under the normal distribution from -1 to 1 is about 0.68, the area under the distribution from -2 to 2 is about 0.95 and the area under the normal distribution from -3 to 3 is nearly 1.0.

On the following page is a table which presents various areas under the normal distribution. The goal of this subsection is to illustrate the use of this table. In particular, we will show how to use this table to compute the area under the normal distribution from -1.5 to 1.5, from 0 to 1.5, from 1.5 to ∞ and from -1.5 to 1.0.

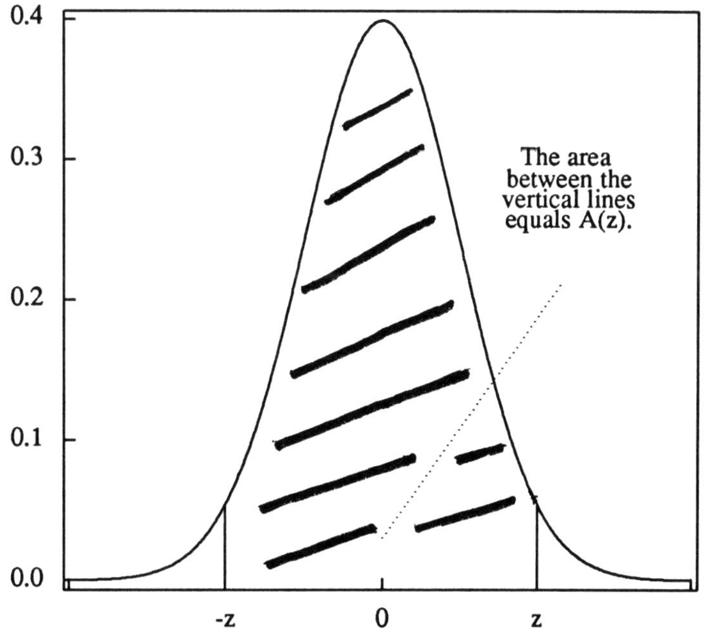

A Normal Table

z	A(z)	z	A(z)	z	A(z)	z	A(z)
0.00	0.0000	0.75	0.5467	1.50	0.8664	2.25	0.9756
0.05	0.0399	0.80	0.5763	1.55	0.8789	2.30	0.9786
0.10	0.0797	0.85	0.6047	1.60	0.8904	2.35	0.9812
0.15	0.1192	0.90	0.6319	1.65	0.9011	2.40	0.9836
0.20	0.1585	0.95	0.6579	1.70	0.9109	2.45	0.9857
0.25	0.1974	1.00	0.6827	1.75	0.9199	2.50	0.9876
0.30	0.2358	1.05	0.7063	1.80	0.9281	2.55	0.9892
0.35	0.2737	1.10	0.7287	1.85	0.9357	2.60	0.9907
0.40	0.3108	1.15	0.7499	1.90	0.9426	2.65	0.9920
0.45	0.3473	1.20	0.7699	1.95	0.9488	2.70	0.9931
0.50	0.3829	1.25	0.7887	2.00	0.9545	2.75	0.9940
0.55	0.4177	1.30	0.8064	2.05	0.9596	2.80	0.9949
0.60	0.4515	1.35	0.8230	2.10	0.9643	2.85	0.9956
0.65	0.4843	1.40	0.8385	2.15	0.9684	2.90	0.9963
0.70	0.5161	1.45	0.8529	2.20	0.9722	2.95	0.9968

Find the area under the normal distribution from -1.5 to 1.5. In general, when asked to calculate an area between two points under the normal distribution, it helps to draw a normal distribution and then notate the area to be computed. In this book, the normal distributions were drawn with a $5000 laser printer. One can do just as well by hand sketching a bell-shaped curve.

To calculate the area between -1.5 and +1.5 under the normal distribution, draw a normal distribution and note the area to be computed:

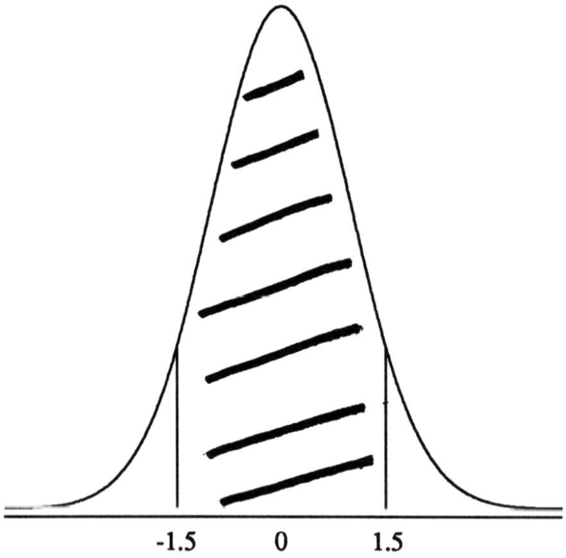

To compute this area, look down the z column of the table, find 1.5 and read off the value of 0.8664. To find the area under the normal distribution from 0.0 to 1.5, divide 0.8664 by two to obtain 0.4332, since the the normal distribution is *symmetric*.

Find the area under the normal distribution from 1.5 to ∞. First sketch the normal distribution and note the area to be computed:

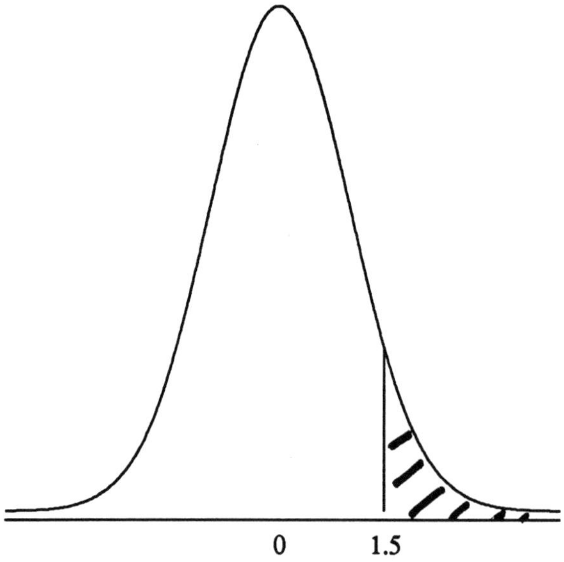

Recall that the area under the normal distribution equals 1.0. From the sketch, the area under the normal distribution from 1.5 to ∞ plus the area under the normal distribution from $-\infty$ to -1.5 equals $1.0 - 0.8664 = 0.1336$. Since the normal distribution is symmetric, the area under the distribution from 1.5 to ∞ is one-half of 0.1336 or 0.0668.

4.5 THE CENTRAL LIMIT THEOREM

To understand what the Central Limit Theorem has to say, consider the following two stage experiment:

(a) THE BASIC n-TOSS EXPERIMENT - Toss the coin (where the probability of heads at any toss is the known value p) n times. Compute the proportion of heads ($\hat{p}_i$) in these n tosses.

(b) Repeat the basic n-toss experiment one hundred thousand times, to obtain $\hat{p}_1, \hat{p}_2, \ldots, \hat{p}_{100,000}$,

The **Central Limit Theorem** states that for *large values of n,* the distribution of the one hundred thousand scaled proportions:

$$\frac{\hat{p}_i - p}{\sqrt{p(1-p)/n}},$$

can be approximated by the normal distribution. The Central Limit Theorem indicates that the average of the one-hundred thousand proportions would be *about p,* that the standard deviation of these proportions would be about

$$\sqrt{p(1-p)/n},$$

and that the histogram of the one hundred thousand scaled proportions would be unimodal, symmetric and more-or-less follow the shape of the normal distribution. For our purposes, n is large if $n(p)(1-p)$ is greater than 5.0.

We can now answer the following type of question:

Assume one tosses a coin, where $p = \frac{1}{2}$, five hundred times and computes the proportion of heads. This basic 500-toss experiment is repeated one hundred thousand times. What proportion of the numbers $\hat{p}_1, \hat{p}_2, \ldots$, and $\hat{p}_{100,000}$ fall in the range [0.476,0.524]?

Since in this example $n(p)(1-p)$ (= 500(0.5)(0.5) = 125) is greater than 5.0, one is safe in using the Central Limit Theorem to answer the question.

To compute the proportion of $\hat{p}_1, \hat{p}_2, \ldots,$ and $\hat{p}_{100,000}$ which lie in the range of [0.476,0.524], draw a rough sketch of the histogram of the one hundred thousand proportions and the area under the histogram of interest:

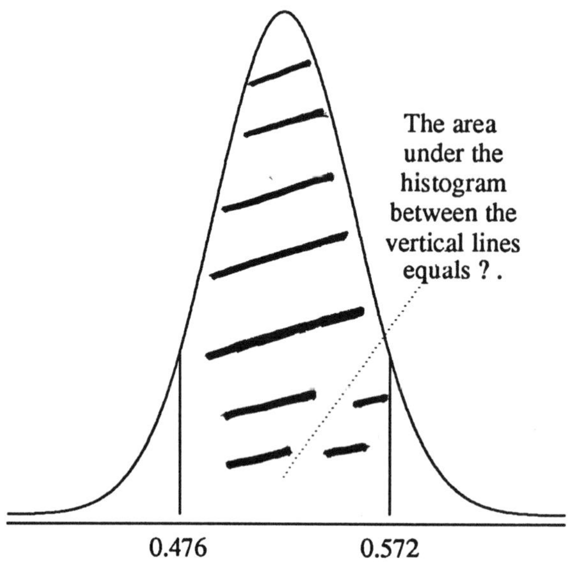

The Central Limit Theorem indicates that the distribution of the *scaled* proportions should follow the normal distribution. For this reason, one scales the endpoints of the interval, i.e. 0.476 and 0.524 using the formula

$$\frac{end\ point - p}{\sqrt{p(1-p)/n}}$$

to compute the area of interest. In particular, scale 0.476 using

$$\frac{(0.476-0.5)}{\sqrt{(0.5 \cdot 0.5)/500}} = -1.07,$$

because for a *fair* coin $p = 0.5$ and in the basic experiment $n = 500$. Similarly, 0.524 is scaled using:

$$\frac{(0.524-0.5)}{\sqrt{(0.5 \cdot 0.5)/500}} = 1.07\ .$$

By appealing to the normal table, the area under the normal distribution from -1.07 to 1.07 is about 0.72.

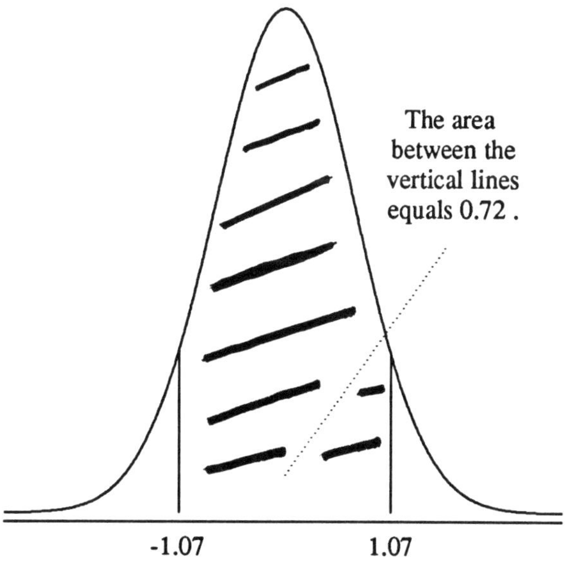

Seventy-two percent of the numbers $\hat{p}_1$, $\hat{p}_2$, ... , and $\hat{p}_{100,000}$ lie between 0.476 and 0.524.

Given that seventy-two percent of the proportions lie between 0.476 and 0.524, the chance that a randomly selected proportion will lie between 0.476 and 0.524 is seventy-two percent. In other words, if a *fair* coin was flipped five hundred times, then there is a seventy-two percent chance that the proportion of heads in the five hundred tosses lies between 0.476 and 0.524. Put yet another way, if a *fair* coin is flipped five hundred times, by the normal approximation, there is a seventy-two percent chance of observing between 238 and 262 heads. In this way, rather than evaluate

$$\binom{500}{238}0.5^{238}0.5^{262} + \binom{500}{239}0.5^{239}0.5^{261} + \binom{500}{240}0.5^{240}0.5^{260} + \ldots + \binom{500}{262}0.5^{262}0.5^{238},$$

to obtain the exact answer, one can use the normal distribution to obtain *an approximate* answer.

4.6 MATERIALS

One coin (preferably new and of the same denomination) for each student and a data sheet.

4.7 EXPERIMENTAL PROTOCOL

NAME______________________

To *FLIP* a coin, rest your thumb between your index and middle fingers. Place the coin on your thumb. Now flick your thumb giving the coin an upward motion. To *SPIN* a coin, balance the coin on edge between a flat surface and the index finger of one hand. Tap the coin off-center with the index finger of the other hand to induce a spinning motion. You will need to apply a sufficiently strong force to the coin so that it develops a good spin. If the coin falls on the floor or if the coin rolls away, repeat the toss or spin. If the coin does not spin for at least four seconds, repeat the spin.

1. **FLIP** the coin thirty (30) times and then **SPIN** the coin thirty (30) times. Record the results on the next page after each spin or flip using "H" or "T".

2. After you have completed the experiment, hand in the following page so that the class results may be pooled.

DATA SHEET

	FLIPPING	SPINNING
1.	________	________
2.	________	________
3.	________	________
4.	________	________
5.	________	________
6.	________	________
7.	________	________
8.	________	________
9.	________	________
10.	________	________
11.	________	________
12.	________	________
13.	________	________
14.	________	________
15.	________	________
16.	________	________
17.	________	________
18.	________	________
19.	________	________
20.	________	________
21.	________	________
22.	________	________
23.	________	________
24.	________	________
25.	________	________
26.	________	________
27.	________	________
28.	________	________
29.	________	________
30.	________	________

Total Number of Heads by *FLIPPING* ________________

Total Number of Heads by *SPINNING* ________________

4.8 ANALYSIS

Your instructor will provide you with the class results so that you may answer the following questions. You will be given a $mx4$ rectangular array of numbers, where the m rows correspond to the m students who participated in the investigation and the four columns correspond to the number of heads obtained by spinning, the number of spins, the number of heads obtained by flipping, and the number of flips, respectively, for the given participant. In questions 4.1 and 4.2, you will investigate to which value the flipping process and the spinning process seem to converge. In questions 4.3-4.9, you will examine the data from the perspective of the Central Limit Theorem. Questions 4.10 and 4.11 address the question of whether flipping and spinning lead to different results. In question 4.12, you will consider two alternative designs of this experiment.

4.1. Each student was asked to flip a coin thirty times. Thus, for each student you should have the number of heads which appeared in thirty flips. Using the class results, compute the cumulative proportion of heads and plot this value versus the total number of flips. For example, if your class consisted of three students and if student #1 had six heads in thirty flips, student #2 had fifteen heads in thirty flips and student #3 had ten heads in thirty flips, then the cumulative proportions would be $\frac{6}{30}$, $\frac{21}{60}$, and $\frac{31}{90}$ and you would plot these proportions versus 30, 60 and 90, respectively. What does your plot reveal regarding the probability of observing heads in the flip of a coin?

4.2. Repeat question 4.1 for the spin data. Compare the spin and the flip plots. How do the plots differ? How are they similar?

4.3. For the flip data, compute each student's proportion of heads. What is the average of these m proportions? How far is the average of these proportions from the probability of heads for a *fair* coin, i.e. 0.5? What is the standard deviation of these m proportions? How far is the standard deviation of these proportions from

$$\sqrt{(0.5 \cdot 0.5)/n} \text{ ?}$$

(Take n to be thirty. Why is it wrong to take n equal to the number of students who participated in the experiment?)

4.4. Repeat question 4.3 for the spin data. Compare the flip and the spin results. Which average is closer to 0.5? Which standard deviation is closer to

$$\sqrt{(0.5 \cdot 0.5)/n} \text{ ?}$$

4.5. Draw a histogram of the proportions computed in question 4.3. What do you see? (Remember Shape, Location and Spread!) Are there any outliers? Is the histogram skewed or is it symmetric? Where is the histogram located? (The default MINITAB histogram may be distorted due to the way MINITAB selects the bins. In particular, some of the bins may contain two distinct proportions, while other bins may only contain one distinct proportion. For this reason, if you use MINITAB, you may have to specify the bin locations and widths for these data.)

4.6. Compute the probability that a *fair* coin will give 0 heads, 1 head, 2 heads, ..., 30 heads in thirty tosses, using the binomial distribution. How do the histograms you obtained in question 4.5 compare with what you would expect under the binomial distribution with $p = 0.5$ and $n = 30$? Given the nature of the flip histogram, does a "fifty-fifty" chance of heads appear reasonable for flipping coins? Repeat these questions for spinning a coin.

4.7. Based on the normal approximation, what is the probability that the proportion of heads in thirty flips of a coin will lie in the interval [0.45, 0.55]? To answer this question, use the scaling formula

$$\frac{(\hat{p}-0.5)}{\sqrt{(0.5 \cdot 0.5)/30}} .$$

Based on this probability, how many of the m proportions would you expect to fall in this interval? How many proportions actually fell into this interval? Is the *observed* number that fell into this interval congruent with the number *expected* under the normal approximation? You were told to scale the data using the value 0.5 for p in the equation

$$\frac{(\hat{p}-p)}{\sqrt{p(1-p)/30}} .$$

Using the value 0.5 to scale the data assumes that the coins were *fair*. Is this assumption reasonable in light of your data?

4.8. If the normal distribution does not seem to provide a reasonable approximation to these data, do you think an n of thirty is sufficiently large for the normal approximation to hold? (What is $np(1-p)$?) Are there any other reasons you can think of to explain a poor performance by the normal approximation? (Think about how the experiment was *implemented*. How could the data get screwed-up?)

4.9. Repeat questions 4.7, and 4.8 for the spin data.

4.10. Based on your responses to questions 4.1 through 4.9, is there any indication that the probability of observing a head differs when a coin is flipped, as opposed to spun?

4.11. To further address the question of the effect of spinning and flipping a coin on the probability of heads, perform the following computation. For each student, record a "1" if *more heads* were observed in *spinning* a coin as in *flipping* a coin. Record a "-1" if *fewer heads* were observed in *spinning* a coin than in *flipping* a coin. Record a "0" if *the same number of heads* were observed in *spinning* a coin as in *flipping* a coin. (For example, if I observed 12 heads in FLIPPING a coin and 15 heads in SPINNING a coin, then I would record a "1".) Compute the number of "1"'s in your data set and denote this total by the letter k. Compute the number of "-1"'s in your data set and denote this total by the letter j. Let $m = j + k$. What we want to determine is whether the value of k which you obtained is unusual. Evaluate the following probability using your values of k and n:

$$\binom{m}{k}0.5^k0.5^{m-k} + \binom{m}{k-1}0.5^{k-1}0.5^{m-k+1} + \binom{m}{k-2}0.5^{k-2}0.5^{m-k+2} + \ldots + \binom{m}{0}0.5^0 0.5^m.$$

What does this quantity represent? This quantity represents the probability of observing k or fewer pairs in which there were *more heads* obtained by *spinning* the coin than in *flipping* the coin, *assuming that the probability of observing a head in a spin of a coin equals the probability of observing a head in a flip of a coin.* Presumably if these two probabilities are equal, then k and j should be about the same. What your probability calculation is going to tell you is whether you observed an unusually *small* value of k. Based on your value for this probability, is there any indication that the probability of observing a head differs when a coin is flipped than when it is spun?

4.12. In this investigation, everyone flipped a coin thirty times and then spun a coin thirty times. Discuss the benefits and drawbacks of such a design as compared to

a. Everyone is *randomized* to flip a coin thirty times *or* to spin a coin thirty times.

b. Everyone is randomized to either flip a coin thirty times and then spin a coin thirty times *or* to spin a coin thirty times and then flip a coin thirty times.

4.9 MINITAB APPENDIX

In MINITAB, it is quite simple to evaluate

$$\binom{n}{k}p^k q^{(n-k)},$$

for any value of n, k and p. The command

BINOmial distribution with n equal to 220 and p equal to 0.32

or more simply BINO 220 .32, will respond with

$$P\,(exactly\ j\ heads) = \binom{220}{j} 0.32^j 0.68^{(220-j)} ,$$

evaluated at j = 0, 1, 2, ... 220, as well as the cumulative probability $P\,(exactly\ 0 +'s)$ + $P\,(exactly\ 1 +)$ + $P\,(exactly\ 2 +'s)$ + ... + $P\,(exactly\ k +'s)$ given by

$$\binom{220}{0} 0.32^0 0.5^{220-0} + \binom{220}{1} 0.5^1 0.5^{219} + \binom{220}{2} 0.5^2 0.5^{218} + \ldots + \binom{220}{k} 0.5^k 0.5^{220-k} ,$$

for $k = 0, 1, 2, \ldots, n$.

The command sequence

```
CDF 1.8;
NORMal 0 1.
```

will compute the area under the normal distribution *to the left of 1.8*. Note that only versions of MINITAB dated release 5.1 or later will have this feature. If you have an older version of MINITAB, then your version of MINITAB will not recognize this command sequence.

To transform data in MINITAB using the square root transformation, use the command

SQRT each value in column C1 and put the transformed values in column C2

or more simply SQRT C1 C2 . This particular command sequence will take the square root of each of the values in C1 and put them in column C2. If a particular value in C1 is negative, then MINITAB will put an asterisk in the corresponding position in C2 and flag the problem with a message. The command LOGE C1 C2, will take natural logarithms of the data in column C1 and put the transformed values in C2. The command DIVIde 1.0 C1 C2, will form the reciprocals of the values in C1 and place the reciprocals in C2. Finally, the command RAISe C1 .3333 C2 will exponentiate each value in C1 to the one-third power (i.e. take the cube root) and put these cube roots in C2.

To simulate the results of tossing a coin 20 times, with the probability of heads at any toss equal to 0.32, use the commands RAND and BINO. In particular, the command draw at

```
RANDom 1 observation and put in column C1;
BINOmial distribution with n = 20 and p = 0.32.
```

or more simply

```
RAND 1 C1;
BINO 20 0.32.
```

will place the simulated number of heads in column C1. If we wanted to repeat five hundred times this basic experiment of tossing a coin 20 times, with the probability of heads at each toss equal to 0.32, then the command draw at

```
RANDom 500 observations and put in column C1;
BINOmial distribution with n = 20 and p = 0.32.
```

or more simply

```
RAND 500 C1;
BINO 20 0.32.
```

will put the five hundred simulated values in column C1. Each simulated value represents the number of heads in twenty tosses of this unfair coin.

To simulate the results of tossing a (six-sided) die, we again use the commands RAND and INTE. In particular, the command draw at

```
RANDom 500 observations and put in column C1;
INTEgers between 1 and 6.
```

or more simply

```
RAND 500 C1;
INTE 1 6.
```

will put the 500 simulated tosses of a die (the 500 numbers that appeared "face up") in column C1.

Simulating data which follow the normal distribution is done via the commands RAND and NORM:

```
RANDom 500 observations and put in column C1;
NORMal 0 1 distribution.
```

or more simply

```
RAND 500 C1;
NORM 0 1.
```

will put 500 fake data points which follow the normal distribution in column C1.

Note that only versions of MINITAB dated release 5.1 or later will have the commands RAND, BINO, INTE and NORM. If you have an older version of MINITAB, then your version of MINITAB will not recognize these commands.

Finally, to answer question 4.1 you will need to know how to compute cumulative proportions. For example, if your class consisted of three students and if student #1 had six heads in thirty flips, student #2 had fifteen heads in thirty flips and student #3 had ten heads in thirty flips, then store the values 6, 15 and 10 in column C1 and the values 30, 30 and 30 in column C2. The command

```
PARSum of the values in column C1 and put in column C10
```

or more simply PARS C1 C10 , will put the partial sums (6, 21 and 31) in column C10. The command

```
PARSum of the values in column C2 and put in column C11
```

or more simply PARS C2 C11 , will put the partial sums (30, 60 and 90) in column C11. To compute the cumulative proportions 6/30, 21/60 just give the command

```
DIVIde the values in column C10 by the values in column C11 and
store the results in column C12
```

or more simply DIVI C10 C11 C12 , and the required proportions will be stored in column C12. We can then PLOT C12 vs C11 to obtain the desired graph.

4.10 REFERENCE

Freedman, D., Pisani, R. and Purves, R. *Statistics,* Norton, New York, 1978.

5

Pies and Needles

Key Words: Statistical Inference; Population; Sample; Estimation; Standard Error; Buffon's Needle Experiment; Binomial Proportion; Normal Distribution.

> For no study is less alluring or more dry and tedious than statistics, unless the mind and imagination are set to work or that person studying is particularly interested in the subject; which is seldom the case with young men in any rank of life.
>
> -William Playfair (1801)

5.1 INTRODUCTION

To illustrate ideas of statistical inference, you will perform an experiment which was first considered some 250 years ago. This experiment has something to do with the notion of measurement. You will be asked to measure the length of a needle. If you had a ruler it would be a simple task - or would it? What if the ruler was graduated in sixteenths of an inch, but you were required to measure the needle to the nearest sixty-fourth of an inch? Would you buy a ruler with a finer graduation? In this chapter, you will perform a version of *Buffon's needle experiment* to illustrate the nontrivial point that one can use "chance" to measure the length of a needle to a precision greater than that of the ruler.

5.2 BUFFON'S NEEDLE EXPERIMENT

Georges Louis Leclerc, Compte de Buffon (1707-1788) left an impact on many branches of science. Among biologists, geologists, philosophers and naturalists, the name Buffon is associated with an extensive treatise entitled *Histoire naturelle, générale et particulière*. This work consists of forty-four volumes, thirty-six of which were written over the period 1749 to 1788, while the remaining eight volumes were published after Buffon's death. It is told that Buffon would work up to twelve hours a day for eight months out of the year, over the forty-year period, in writing this work. As the name suggests, the *Histoire naturelle, générale et particulière* is an attempt to produce an account of the whole of nature. The work consists of descriptions of animals, reptiles, and fish, as well as wonderful color prints relating to the subject matter and general philosophic discussions on nature.

To probabilists and mathematicians, Buffon is remembered for his "experiment" in geometric probability. In the essay, "Essai D'Arithmétique Morale" first published in 1777, Buffon posed and solved a curious problem: Assume one has a needle of length a and a grid of parallel lines with common distance d ($a \le d$) between them (Buffon imagined this grid to be a floor with parallel lines as a design), something like:

The Set-Up for the Buffon Experiment

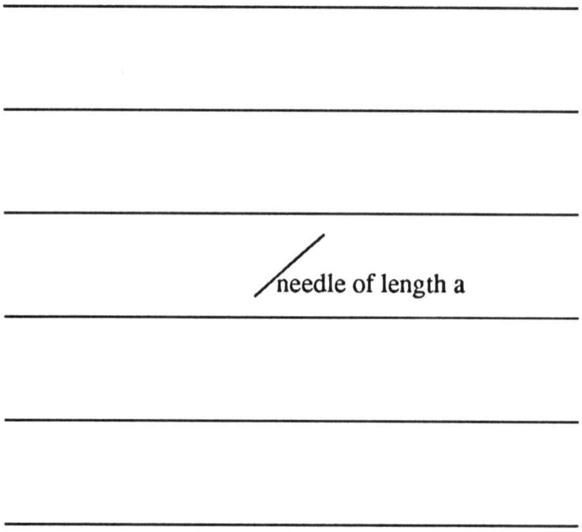

What is the probability (p) that a randomly tossed needle will cross one of the lines? Buffon showed, in a mathematical derivation, that this probability is about two-thirds the ratio of the length of the needle to the inter-line distance. More formally, Buffon found that the probability of interest is given as:

$$p = \frac{2a}{\pi d},$$

where $\pi = 3.14159...$, that is twice the length of the needle divided by a little more than 3.14 times the interline distance.

The mathematical derivation of this result will not be presented. "Essai d'arithmétique morale" can be found in the collected works of Buffon entitled *Oeuvres Complètes de Buffon* (1853-55). Uspensky (1937) or Gnedenko (1962) present a mathematical derivation of this result.

Regarding this formula, note two important points. First, the probability that the needle will cross a line increases linearly with the length of the needle. Second, the probability that the needle will cross a line is inversely related to the distance between the lines. Clearly, these two points are not unexpected. As the length of the needle is increased, one would expect the probability of the needle to cross a line to increase. Buffon's formula indicates that this increase in probability is linearly related to the length of the needle. Similarly, as one moves the lines further apart, one would expect the probability of the needle to cross a line to decrease. Buffon's formula indicates that the decrease is proportional to the inverse of the distance between the lines.

To obtain a rough idea of the length of the needle, one need only compare the needle to the distance between the parallel lines on Buffon's floor. By making such a comparison, one can learn that the length of the needle is less than d units. But what if one requires a more precise estimate of the length of the needle? The idea is to rely on the *planned introduction of chance* to *estimate* the length of the needle.

The statistical procedure consists of tossing a needle at random onto a floor with parallel lines and then to compute the proportion of trials in which the needle crosses a line. This proportion ($\hat{p}$) will be the estimate of the true probability that a randomly tossed needle will cross a line. Now set this observed proportion equal to the formula for the probability that the needle will cross a line:

$$\hat{p} = \frac{2a}{\pi d} .$$

In this way, if one has a value for $\hat{p}$, if one knows the numerical value for π and if one knows d, then one can solve this equation for $\hat{a}$, the estimated length of the needle:

$$\hat{a} = \frac{\hat{p}\pi d}{2} .$$

Thus, one can "plug in" $\hat{p}$, π and d to obtain an estimate of the length of the needle.

What assures us that this procedure will give a sufficiently precise answer? According to the Frequentists, (see Section 4.2) if one is willing to do enough tosses, then $\hat{p}$ will be a very precise estimate of p. Therefore, because one has a procedure to determine p to an arbitrary degree of accuracy, it follows that the length of the needle (a) can be determined to any level of precision.

Assume the needle is tossed n times. How large a difference between $\hat{a}$ and a should one expect? The answer to this question is related to the idea of the standard error of the estimate $\hat{a}$.

5.3 THE STANDARD ERROR OF $\hat{a}$

How large a difference between $\hat{a}$ and a should one expect? The answer to this question is given by the *standard error* of $\hat{a}$. The goal of this section is to present the standard error of $\hat{a}$.

Assume that one tosses a needle n times onto a floor with parallel lines and then repeats this basic experiment one hundred thousand times. Given the results of these one hundred thousand repetitions of the basic experiment, one would be able to construct the distribution of $\hat{a}_1, \hat{a}_2, \ldots., \hat{a}_{100000}$ for the particular needle and for the particular floor, that is, for the particular values of a and d. The average of these one hundred thousand $\hat{a}_i$'s would be about a. It is of interest to develop a formula for the standard deviation of the one hundred thousand $\hat{a}_i$'s.

If n is *sufficiently large,* then the histogram of the one hundred thousand $\hat{a}_i$'s would follow the normal distribution. In this way, one would expect about 68% of the $\hat{a}_i$'s to fall within one SD of the true value a and about 95% of the $\hat{a}_i$'s to fall within two SD's of a. If the SD of the one hundred thousand $\hat{a}_i$'s is small, then this would imply that the $\hat{a}_i$'s are tightly concentrated about a. If the standard deviation of the one hundred thousand $\hat{a}_i$'s is not small, then this would imply that the $\hat{a}_i$'s are somewhat more dispersed about a.

It can be shown that the standard deviation of the $\hat{a}_i$'s is given by

$$\sqrt{\left(\frac{\pi d}{2}-a\right)\frac{a}{n}} \; .$$

In other words, among the one hundred thousand $\hat{a}_i$'s, the *typical* $\hat{a}_i$ is about

$$\sqrt{\left(\frac{\pi d}{2}-a\right)\frac{a}{n}}$$

units from a.

Given a single sample of size n, how far would one expect $\hat{a}$ to be from a? One would expect $\hat{a}$ to be the typical distance from a, i.e.

$$\sqrt{\left(\frac{\pi d}{2}-a\right)\frac{a}{n}} \; .$$

Of course, since one does not have a value for a, $\hat{a}$ is plugged into the standard error formula to determine how far one expects $\hat{a}$ to be from a. Thus, if the needle is randomly tossed n times onto the single grid, then the estimate of the length of the needle ($\hat{a}$) is expected to differ from the true value (a) by about

$$\sqrt{\left(\frac{\pi d}{2}-\hat{a}\right)\frac{\hat{a}}{n}} .$$

This formula represents the estimated standard error of the estimate $\hat{a}$.

5.4 MATERIALS

Each student should have a "needle", a data sheet and a grid.

5.5 EXPERIMENTAL PROTOCOL

1. Place the grid on the floor. (Smooth out the wrinkles!)

2. Stand two feet away from the grid.

3. Toss the "needle" onto the grid. If the needle misses the grid, then try again! In case the "needle" touches a line (without crossing it), repeat the toss.

4. Place an 'x' on the data sheet if the "needle" CROSSES a horizontal line. Place an 'o' on the data sheet if the "needle" DOES NOT CROSS a horizontal line.

5. Repeat steps 2-4 thirty times.

6. Based on *your* results fill in the following blank:

My needle crossed a line __________ times out of thirty trials.

DATA SHEET

NAME____________________

1. __________
2. __________
3. __________
4. __________
5. __________
6. __________
7. __________
8. __________
9. __________
10. __________
11. __________
12. __________
13. __________
14. __________
15. __________
16. __________
17. __________
18. __________
19. __________
20. __________
21. __________
22. __________
23. __________
24. __________
25. __________
26. __________
27. __________
28. __________
29. __________
30. __________

Total # of crosses __________

Hand in this page, so that the data may be pooled. Remember to record *your* results on the previous page before you hand in this sheet.

5.6 ANALYSIS

To answer question 5.1 you will require the results of your experiment. To answer questions 5.2-5.6, you will require the results of the k students in your class who completed this investigation. Your instructor will provide you with the class results, as well as the "true" value of a. You will receive a column of k numbers, where k is the number of students that participated in the experiment and each number represents the number of times the "needle" crossed the line, out of thirty trials, for a given student. Question 5.1 relates to your results, while questions 5.2 to 5.6 relate to the class results. Question 5.2 looks at the problem from a Frequentist point of view, while questions 5.3-5.6 look at the problem from a Central Limit Theorem point of view. Question 5.7 raises the possibility of subtle biases in the implementation of this experiment.

5.1. Based on your data, compute $\hat{a}$ and its *estimated* standard error. How far off was *your* estimate of the length of the needle from the true length? How far off would you have *expected* your estimate to have been in light of its *estimated* standard error?

5.2. Based on the results of the k students in the class who completed this investigation, compute the following cumulative probability:

$$\hat{p} = \frac{\textit{number of crosses of a line in m tosses}}{m},$$

for m = 30, 60, ... 30k. Plot $\hat{p}$ versus m. For example, if there were three students in your class and these students obtained values of $\frac{12}{30}$, $\frac{9}{30}$ and $\frac{10}{30}$, then you would plot $\frac{12}{30}$ versus 30, $\frac{21}{60}$ versus 60 and $\frac{31}{90}$ versus 90. Describe what you see in your cumulative plot of $\hat{p}$.

5.3. Draw a histogram of the $\hat{a}$'s. (Note that the default MINITAB histogram may give a distorted view of the distribution of the data due to its selection of the bin width. You may have to select an alternative bin width to avoid this bias.) Describe the shape of the histogram. Is it symmetric? Is it unimodal? Does the histogram appear to be centered about any value? Are there any outlying values? Can you determine what caused these values to be peculiar? Are there any peculiarities to the shape of the histogram?

5.4. Compute the average of the k $\hat{a}$ values (call this number ave($\hat{a}$)). What is ave($\hat{a}$) - a, that is, on average how far off were the students in your class in measuring the length of the needle? How do any outlying $\hat{a}$ values effect ave($\hat{a}$)?

5.5. For each of the k $\hat{a}$'s, compute its *estimated* standard error. Compute the average of the k *estimated* standard errors of $\hat{a}$. Compare this number to the standard deviation of the k $\hat{a}$ values. (Why should these two numbers be similar in value?)

5.6. Refer to the histogram in 5.3. Is there any evidence of a systematic departure from a "bell" shape for the distribution of the $\hat{a}$'s? If there are any such departures, how would you explain these departures? Do you think there were any problems in the *implementation* of the experiment which would give rise to such departures? Do you think that $n = 30$ is too small for the Central Limit Theorem to take effect?

5.7. What aspects of the experiment can affect the *quality* of the data? Discuss each of the following components of the experiment: the "needles", the grids, the tosses and the recording process. What would you do to eliminate any sources of error?

5.7 REFERENCES

1. Gnedenko, B. V. *Theory of Probability,* Chelsea, New York, 1962.

2. *Oeuvres Complètes de Buffon,* edited by M. Flourens, Garnier Frères, Paris, vol. 12, [1853-1855].

3. Uspensky, J. V. *Introduction to Mathematical Probability,* McGraw-Hill, New York, 1937.

6

What's the Answer? What's the Question?

Key Words: Sample Surveys; Evasive Answer Bias; Randomized Response; Binomial Proportion; Standard Error; Confidence Interval; Simple Random Sample.

How far would Moses have gone if he had taken a poll in Egypt?
-Harry S. Truman

6.1 INTRODUCTION

The purpose of this chapter is to discuss a problem which can arise in the implementation of a survey, thereby, distorting the results of the research and one approach for dealing with this problem. The problem on which we will focus, arises when the participants in a survey are questioned about sensitive or embarrassing issues. In such situations some people may refuse to respond to such questions or may give false information; that is, they may respond in a manner which they feel is closer to the "norm" rather than respond according to their feelings or behavior.

As an example, to understand the dynamics of the AIDS epidemic it is important to study sexual behavior among adults in the United States. However, to do so would require asking people some very sensitive questions. For example, one may ask a man whether he has had sex with a prostitute during the past month. Clearly, this question treats a topic which some men may deny. As such, if one were to simply ask this question, the proportion of men who answer "yes" is expected to *underestimate* the proportion of men in the population of interest who actually had sex with a prostitute during the past month. This problem of **evasive answer bias** and one approach for dealing with this issue are the focus of this chapter. In particular, the technique of *randomized response* will be discussed. This technique protects the privacy of the individual, yet still attempts to provide useful survey information. You will then implement this technique with a survey of the members of your class.

6.2 THE TECHNIQUE OF RANDOMIZED RESPONSE

How should a sensitive question be handled in a survey? Warwick and Lininger (1975) suggest that the *placement* of the question is very important. Campbell (1987) suggests that the *presentation* of the sensitive question may also help to overcome the respondents' reluctance to accurately answer the sensitive questions. In particular, Campbell suggests that the participant be handed a card which shows the response categories, each category being associated with a letter (a, b, c, ...) The respondent then answers with a letter, rather than having to describe the particular behavior. By not being forced to vocalize the behavior, Campbell argues that the person will be more willing to respond accurately.

While the placement of the sensitive question and the manner in which the sensitive question is presented are important, another approach can be used to address the problem of sensitive questions.

The statistical technique of **randomized response** may reduce or even eliminate the problem of evasive answer bias. Each participant in the survey is asked to flip a *fair* coin and *not* disclose the result of the toss to the interviewer. The interviewer then asks the sensitive question (such as "Have you had sex with a prostitute during the past month?"). The participant is then instructed to answer "no" if the coin came up tails *and* the participant has *not had* sex with a prostitute during the past month. Otherwise, the participant should answer "yes". In this way, "no" means

> The coin came up tails *and* the respondent has not had sex with a prostitute during the last month.

While a response of "yes" means

> The coin came up tails *and* the respondent did have sex with a prostitute during the last month,

or

> The coin came up heads *and* the respondent has not had sex with a prostitute during the last month,

or

> The coin came up heads *and* the respondent did have sex with a prostitute during the last month.

Only the participant knows whether the "yes" response reflects the result of the coin toss or the participant's experience. Thus, it is believed that the participant may be less inclined to falsify the response or refuse to respond.

After interviewing one hundred people one may have sixty affirmative replies and forty negative replies. How is one to use this information to estimate the proportion of men who had sex with a prostitute during the last month in the target population? The success of the randomized response technique in teasing-out the required information is based on the following fact: About 50% of those who did not have sex with a prostitute during the past month would obtain a tail on their coin toss. Thus, whatever proportion answered "no", one simply doubles that proportion to obtain an estimate of the proportion of men in the target population who did not have sex with a prostitute during the previous month. (For future reference, we will refer to the proportion in the sample who respond with a "no" as $\hat{\rho}$ and we will let $\hat{P} = 2\hat{\rho}$ denote our estimate of the proportion of men in the target population who did not have sex with a prostitute in the last month.) In this way, the estimate of the proportion of men in the target population who did have sex with a prostitute is $1 - \hat{P}$.

To illustrate randomized response, assume a sample of forty-nine students was drawn from the population of undergraduate students at the University of Wisconsin to participate in the randomized response experiment of Section 6.7. Of these forty-nine students, twenty students answered "no" regarding the tonsillectomy question. In this way, $\hat{\rho} = \frac{20}{49} = 0.41$ and $\hat{P} = 2 \, x \, 0.41 = 0.82$. Thus, it is estimated that eighty-two percent of the undergraduate students at the University of Wisconsin *do not* have the "sensitive" attribute, while eighteen percent do have the "sensitive" attribute of having had a tonsillectomy.

How does this randomized response estimate of eighteen percent compare with the "direct" estimate? (By "direct" estimate, it is meant the actual proportion of students in the sample with the sensitive attribute.) Since having had a tonsillectomy carries little social stigma, in addition to playing randomized response, the students were also asked to indicate whether or not they had a tonsillectomy. Twelve of the forty-nine students or about twenty-five percent indicated that they did have a tonsillectomy. Thus, playing randomized response, rather than simply asking the students, induced a slight distortion of the estimate. Of course, the value of randomized response is in situations where the question is sensitive. In such cases, it is assumed that directly asking the question will result in an extremely biased estimate, due to evasive answer bias. However, it is hoped that playing randomized response will help to reduce or even eliminate this bias.

As a historical note, the technique of randomized response was first introduced by Warner (1965).

6.3 THE STANDARD ERROR OF $\hat{P}$

It is important when reporting $\hat{P}$ or $1 - \hat{P}$ to assign a measure of precision to these estimates. Is the proportion in the population well-determined by $\hat{P}$ or is it poorly determined? In other words, would one expect $\hat{P}$ to differ from the true proportion by 3% or 30%?

Suppose that one draws a sample of size *n* WITHOUT REPLACEMENT from a population containing *m* people and with this sample plays randomized response.

Assuming that the only error in the data is due to sampling variability, the estimated standard error of $\hat{P}$ or $1 - \hat{P}$ is given by

$$2\sqrt{\frac{\hat{\rho}(1-\hat{\rho})}{n}\left[\frac{m-n}{m-1}\right]}$$

where $\hat{\rho}$ is the proportion of the sample who respond "no".

To illustrate the computation of the estimated standard error of $1 - \hat{P}$, assume that the sample of forty-nine students was drawn WITHOUT REPLACEMENT from the population of University of Wisconsin undergraduate students (of which in the Fall Semester of 1987 there were 29,777). Recall, that $\hat{\rho}$ is equal to $\frac{20}{49}$, while $\hat{P}$ (the proportion *without* the sensitive attribute) is equal to $2 \times \frac{20}{49} = 0.82$. Thus, the proportion of University of Wisconsin undergraduate students estimated to have had a tonsillectomy is eighteen percent. The estimated standard error of $1 - \hat{P}$, is then

$$2\sqrt{\frac{0.41(1-0.41)}{49}\left[\frac{29777-49}{29777-1}\right]}$$

which equals $2 \times 0.070 \times 0.999 = 0.14$. Thus, we estimate that eighteen percent give or take about fourteen percent of the undergraduates at the University of Wisconsin have had a tonsillectomy.

As pointed out in the previous section, if one *directly* asked the students whether or not they had a tonsillectomy, one would have concluded that about twenty-five percent of the undergraduates at the University of Wisconsin have had a tonsillectomy. The estimated standard error of this direct estimate is

$$\sqrt{\frac{0.25(1-0.25)}{49}\left[\frac{29777-49}{29777-1}\right]}$$

or $0.062 \times 0.999 = 0.062$. Thus, assuming that the sample was drawn WITHOUT REPLACEMENT and that the only error in the data is due to sampling variability, the "direct" approach indicates that one can expect twenty-five percent of the University of Wisconsin undergraduates give or take six percent to have had a tonsillectomy.

Note that the standard error of the estimate of the proportion of the population based on randomized response is about 2.3 times larger ($0.14/0.06 = 2.3333...$), than the corresponding standard error for the "direct" estimate. By flipping the coin in randomized response one adds to the chance error of the survey and thereby *increases* the variability of the estimate. Of course, one flips the coin in randomized response to *decrease* the amount of evasive answer bias in the estimate. Thus, there is an inherent conflict between decreasing the bias in the estimate and increasing the variability of the estimate.

In playing randomized response, it is hoped that the drop in bias is sufficient to compensate for the increase in variability of the estimate, for a given sample size.

6.4 CONFIDENCE INTERVAL

In the previous section, a formula for the standard error of $1 - \hat{P}$ was derived. Thus, having conducted a survey of our target population, one can conclude that the proportion of University of Wisconsin undergraduate students who had a tonsillectomy is $1 - \hat{P}$ *give or take*

$$2\sqrt{\frac{\hat{\rho}(1-\hat{\rho})}{n}\left[\frac{m-n}{m-1}\right]},$$

where $\hat{\rho}$ is the proportion of the sample who responded with a "no". If this give or take is small, (for example, 1% or 2%), then one can say that the proportion in the population $1 - P$ is well-determined by $1 - \hat{P}$. On the other hand, if this give or take is not so small, (for example, 15% or 20%) then $1 - \hat{P}$ is a poor estimate of the true population proportion.

In fact, one can be more quantitative. If one forms the interval given by

$$1 - \hat{P} - 2\sqrt{\frac{\hat{\rho}(1-\hat{\rho})}{n}\left[\frac{m-n}{m-1}\right]}, 1 - \hat{P} + 2\sqrt{\frac{\hat{\rho}(1-\hat{\rho})}{n}\left[\frac{m-n}{m-1}\right]}),$$

then it turns out that this interval has a 68% chance of including the value $1 - P$ - the true proportion of University of Wisconsin undergraduate students who have had a tonsillectomy, assuming that the only source of error in the sample is due to sampling variability and that the sample was drawn WITHOUT REPLACEMENT. In general, the **100α% confidence interval** given by

$$(\hat{P} - 2z\sqrt{\frac{\hat{\rho}(1-\hat{\rho})}{n}\left[\frac{m-n}{m-1}\right]}, \hat{P} + 2z\sqrt{\frac{\hat{\rho}(1-\hat{\rho})}{n}\left[\frac{m-n}{m-1}\right]}),$$

where z is the value such that the area under the normal distribution from $-z$ to $+z$ is equal to 100α%, has a 100α% chance of including the proportion in the population (P) which does not have the sensitive attribute, for n *sufficiently large*. Similarly, the 100α% *confidence interval* given by

$$1 - \hat{P} - 2z\sqrt{\frac{\hat{\rho}(1-\hat{\rho})}{n}\left[\frac{m-n}{m-1}\right]}, 1 - \hat{P} + 2z\sqrt{\frac{\hat{\rho}(1-\hat{\rho})}{n}\left[\frac{m-n}{m-1}\right]})$$

has a 100α% chance of including the proportion in the population $(1-P)$ which does have the sensitive attribute, for *n sufficiently large.*

To illustrate the computations, recall that based on our implementation of randomized response, we estimated that eighteen percent of the undergraduate students at the University of Wisconsin had a tonsillectomy. The estimated standard error of this estimate is equal to 0.14. Thus, a 95% confidence interval for the "true" percentage of University of Wisconsin undergraduate students who had a tonsillectomy is

$$(0.18 - 1.96 \cdot 0.14, 0.18 + 1.96 \cdot 0.14)$$

or (0.0,0.45). In this way, one is 95% confident that between zero and forty-five percent of the University of Wisconsin undergraduate students had a tonsillectomy. Clearly, the "true" proportion is not at all well-determined by this estimate.

6.5 MATERIALS

One coin (preferably new and of the same denomination) and a data sheet.

6.6 EXPERIMENTAL PROTOCOL NAME______________________

DATA SHEET

1. Flip the coin. Record the result.

------- **HEAD** ------- **TAIL**

2. If the toss resulted in a **TAIL and** you **DID NOT** have a tonsillectomy, then answer **NO.** Otherwise, answer **YES.**

------- **YES** ------- **NO**

3. For the purpose of comparison, *all* participants should answer the following question.

Did you ever have a tonsillectomy?

------- **YES** ------- **NO**

4. Flip the coin again. *Remember* the result of the coin toss. **DO NOT** record the result of the coin toss.

5. If the toss resulted in a **TAIL and** you **DID NOT** do cocaine during the last SIX months, then answer **NO.** Otherwise, answer **YES.**

------- **YES** ------- **NO**

Hand in this page, so that the class results may be pooled.

6.7 ANALYSIS

Your instructor will provide you with the class results so that you may answer the following questions. In particular, you will be told the total number of students who participated in the investigation, the number who responded with a "no" to the tonsillectomy question and the number who responded with a "no" to the cocaine question.

6.1. Based on the class results, compute the proportion of respondents answering "no" to question 3. Use this result to compute $1 - \hat{P}$. Explain in words what you mean by $1 - \hat{P}$.

6.2. What proportion of the respondents *actually had* a tonsillectomy? How far off was your randomized response based estimate from the "direct" estimate?

6.3. Is it reasonable to conclude that the answers to question 6.2 can be generalized to all the students on your campus? Why? Why not?

6.4. Compute the *estimated* standard error of $1 - \hat{P}$ (*ESE of* $1-\hat{P}$) using the results of your survey. Compute the 68% and 95% confidence intervals for $1 - P$. You may assume that the square root of $\frac{m-n}{m-1}$ is nearly 1.0.

6.5. What assumption have you made to justify that the estimated standard error you obtained in question 6.4 quantifies the *sampling variability* in $1 - \hat{P}$? Is this assumption reasonable? (Hint: Think about how the sample was *obtained.)* Is there any reason to believe that *sampling variability* is not the only source of error in the survey?

6.6. Assuming that the respondents are willing to respond honestly to the sensitive question, explain why the standard error of $1 - \hat{P}$ is larger than the standard error of the proportion in the sample who respond "yes" to the sensitive question.

6.7. If you wanted to follow-up on this experiment to obtain a more precise estimate of $1 - P$, about how many people would you need to sample to insure that the standard error of $1 - \hat{P}$ is under three percent? (Hint: set

$$2\sqrt{\frac{\hat{p}(1-\hat{p})}{n}}$$

equal to 0.03 and solve for *n*.)

6.8. Based on the class results, compute the proportion of respondents answering "no" to the cocaine question. Use this result to compute $1 - \hat{P}$. Explain what you mean by $1 - \hat{P}$.

6.9. Is it reasonable to conclude that your response to question 6.8 can be generalized to all the students on your campus? Why? Why not?

6.10. Compute the *estimated* standard error of $1 - \hat{P}$ (*ESE of* $1-\hat{P}$) using the results of your survey. Compute a 95% confidence interval for $1 - P$. You may assume that the square root of $\frac{m-n}{m-1}$ is nearly 1.0.

6.11. What assumption have you made to justify that the estimated standard error you obtained in question 6.10 quantifies the *sampling variability* in $1 - \hat{P}$? Is this assumption reasonable? (Hint: Think about how the sample was *obtained.)* Is there any reason to believe that *sampling variability* is not the only source of error in the survey?

6.8 MINITAB APPENDIX

Given one hundred tickets in a hat, how can we draw five tickets WITHOUT REPLACEMENT using MINITAB? If the tickets are numbered 1 to 100, and these 100 numbers are stored in column C1, then the following command will select five numbers at random WITHOUT REPLACEMENT between 1 and 100, inclusive:

```
SAMPle 5 observations without replacement from column C1 and put in C2
```

or more simply SAMP 5 C1 C2 . The five numbers that are stored in column C2 are the numbers of the tickets which have been selected at random WITHOUT REPLACEMENT.

6.9 REFERENCES

1. Campbell, A. A. "Randomized Response Technique." *Science, 236,* 1049, 1987.

2. Warner, S. L. "Randomized Response: A Survey Technique for Eliminating Evasive Answer Bias." *Journal of the American Statistical Association, 60,* 63-69, 1965.

4. Warwick, D. P. and Lininger, C. A. *The Sample Survey: Theory and Practice,* McGraw Hill, New York, 1975.

7

Splitting Hairs

Key Words: Confidence Interval; Standard Error; t Distributions; One-Sample Problem; Measurement Error.

He uses statistics as a drunken man uses lamp posts - for support rather than illumination

-Andrew Lang

7.1 INTRODUCTION

This chapter will focus on confidence intervals for the population average (μ). In the next section, the standard error of the average will be introduced. Later in this chapter, a confidence interval for the population average will be presented. To illustrate these notions, you will quantify the error that is incurred when one bisects a line free-hand. You will attempt to estimate the "typical" magnitude of such errors and then produce a range of plausible values for this "typical" value.

7.2 THE AVERAGE AND ITS STANDARD ERROR

The type of statistical problem that will be considered in this chapter is that of the **one-sample problem.** Assume that there is one population of interest and that one draws a sample from this population. Based on the information in this sample, the population average (μ) is to be estimated. To this end, the sample average ($\bar{x}$) will be used to estimate the population average.

In addition, one will want to attach a *measure of precision* to this estimate and to specify a range of plausible values for the population average. The precision of the estimate is given by the standard error of the average. The standard error of the average indicates how far, on average, one would expect $\bar{x}$ (the sample average) to be from μ (the population average).

To motivate the standard error of the average, consider the following sampling experiment. From a hat of tickets let's draw repeated random samples WITH REPLACEMENT of size n. For each of these samples, compute the sample average of the values on the tickets to obtain $\hat{a}_1, \hat{a}_2, \ldots$. When the sample size n is sufficiently large, the histogram of these $\hat{a}_i$'s is symmetric, unimodal and centered at μ. In addition, the standard deviation of $\hat{a}_1, \hat{a}_2, \ldots$ is equal to $\frac{\sigma}{\sqrt{n}}$, where σ is the *modified* standard deviation of the values written on the tickets *in the hat*. In this way, the "typical" $\hat{a}_i$ is $\frac{\sigma}{\sqrt{n}}$ units from μ.

Given the results of one sample, the formula $\frac{\sigma}{\sqrt{n}}$ will be used as a guide to indicate how far one can expect the sample average ($\bar{x}$) to differ from μ. If this quantity is small, then one would expect $\bar{x}$ to be close to μ (i.e. μ is well determined), while if this value is not so small, then $\bar{x}$ may be far from μ. The quantity, $\frac{\sigma}{\sqrt{n}}$, will be referred to as the **standard error of the average.**

In practice, one typically does not know σ. As such, one cannot evaluate the formula $\frac{\sigma}{\sqrt{n}}$ to obtain the standard error of the average. However, given a single sample of observations, one can use the sample standard deviation (the standard deviation of the values on the tickets in the sample) as a proxy for σ. Thus, when one does not know σ, one can estimate the standard error of the average using the formula $\frac{SD}{\sqrt{n}}$.

It is pointed out that it is very important to understand the distinction between the standard deviation (SD) and the standard error (SE). The *standard deviation* relates to a list of numbers and indicates how far, on average, the numbers in the list are from the average of the list. The *standard error of the average* relates to the results of a random sample from a population, in particular to the sample average. The standard error indicates how far, on average, one would expect the sample average to be from the population average.

7.3 THE *t* DISTRIBUTIONS

Before proceeding with a discussion of confidence intervals, the t distributions will be introduced.

Mathematically speaking, the **t distributions** are a family of distributions indexed by a positive integer parameter called the *degrees of freedom*. Thus, there is a t distribution for 1 degree of freedom, for 2 degrees of freedom, ... On the following page, the t distributions for $d = 1$ and $d = 15$ are plotted. As one can see, these t distributions are unimodal, centered at zero, and symmetric. In fact, all t distributions are unimodal, centered at zero, and symmetric. In addition, it turns out that the area under each of the t distributions equals 1.0.

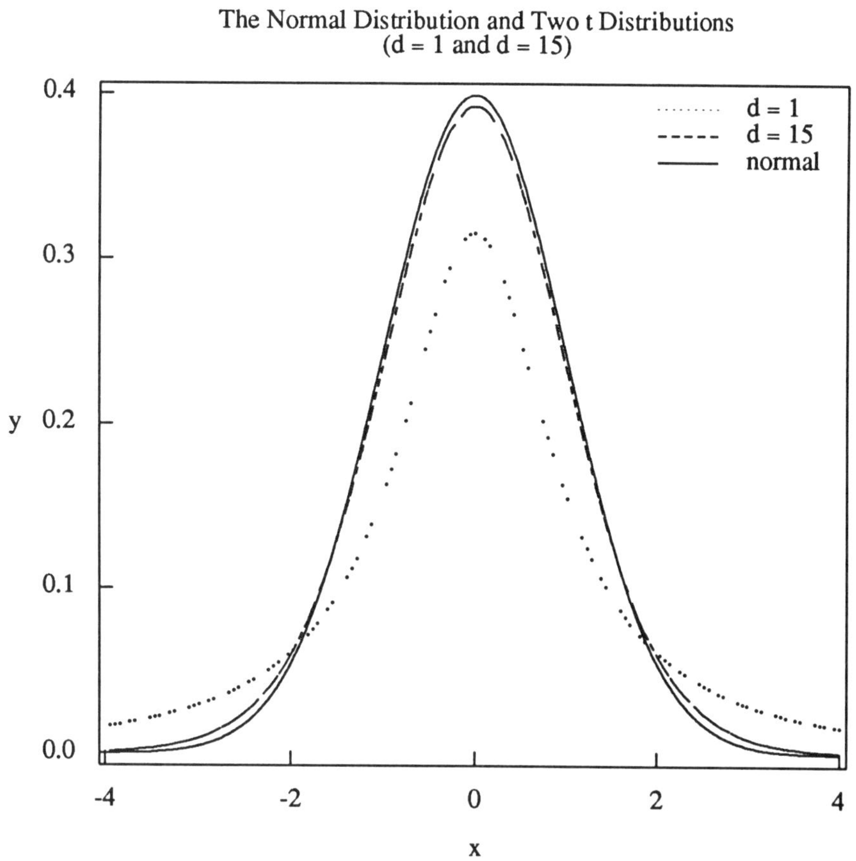

Notice that for small degrees of freedom (look at the distribution corresponding to $d = 1$ degree of freedom), the t distribution is much more dispersed than the normal distribution. However, with a larger degree of freedom, the t distribution looks more normal (look at the distribution corresponding to $d = 15$ degrees of freedom).

The following table presents various areas under selected t distributions. The present goal is to illustrate the use of this table.

Areas Under Various t Distributions

df	0.80	0.90	0.95	0.98	0.99
1	3.078	6.314	12.706	31.821	63.657
2	1.886	2.920	4.303	6.965	9.925
3	1.638	2.353	3.182	4.541	5.841
4	1.533	2.132	2.776	3.747	4.604
5	1.476	2.015	2.571	3.356	4.032
6	1.440	1.943	2.447	3.143	3.707
7	1.415	1.895	2.365	2.998	3.499
8	1.397	1.860	2.306	2.896	3.355
9	1.383	1.833	2.262	2.821	3.250
10	1.372	1.812	2.228	2.764	3.169
11	1.363	1.796	2.201	2.718	3.106
12	1.356	1.782	2.179	2.681	3.055
13	1.350	1.771	2.160	2.650	3.012
14	1.345	1.761	2.145	2.624	2.977
15	1.341	1.753	2.131	2.602	2.947
16	1.337	1.746	2.120	2.583	2.921
17	1.333	1.740	2.110	2.567	2.898
18	1.330	1.734	2.101	2.552	2.878
19	1.328	1.729	2.093	2.539	2.861
20	1.325	1.725	2.086	2.528	2.845
21	1.323	1.721	2.080	2.518	2.831
22	1.321	1.717	2.074	2.508	2.819
23	1.319	1.714	2.069	2.500	2.807
24	1.318	1.711	2.064	2.492	2.797
25	1.316	1.708	2.060	2.485	2.787
30	1.310	1.697	2.042	2.457	2.750
40	1.303	1.684	2.021	2.423	2.704
50	1.299	1.676	2.009	2.403	2.678
60	1.296	1.671	2.000	2.390	2.660
80	1.292	1.664	1.990	2.374	2.639
Normal	1.282	1.645	1.960	2.326	2.576

To understand how this t table works, notice that the *rows* of the table correspond to *degrees of freedom* and the *columns* of the table correspond to *areas* under the various t distributions. A given value at row $df = d$ and under the column labelled $0.x$ gives the value, t, such that between the value $-t$ and the value $+t$ lies *x% of the area under the t distribution on d degrees of freedom.* Notice that when the degrees of freedom are large, the areas are similar to the corresponding areas under the normal distribution.

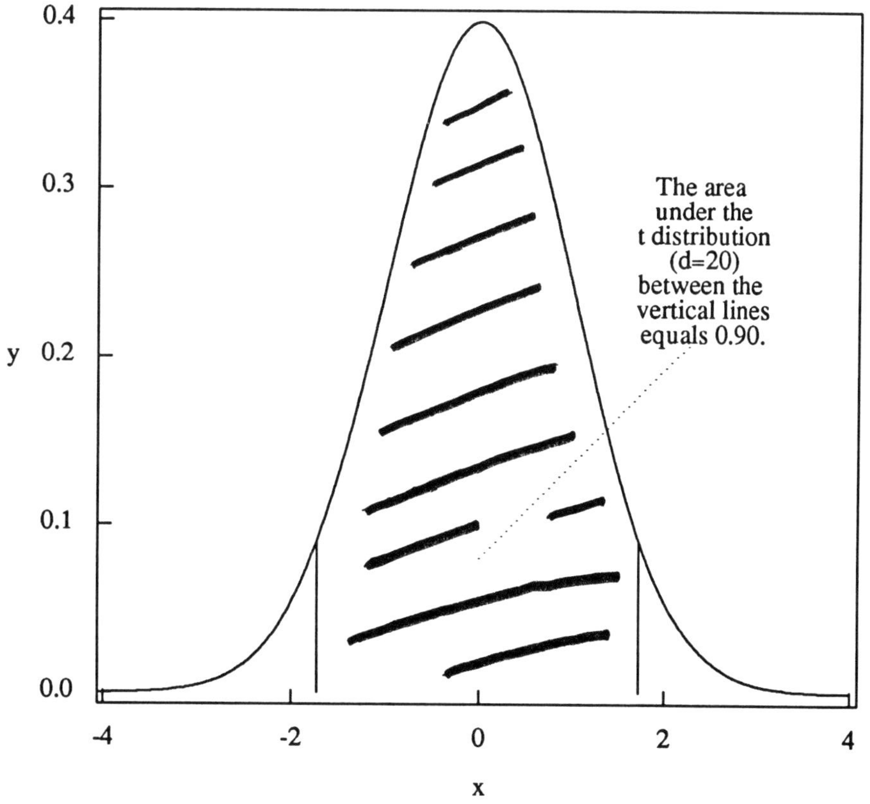

To illustrate the use of this table, consider the following examples:

a. Let's use this table to find the value t such that 90% of the area under the t distribution on twenty degrees of freedom lies between $-t$ and $+t$. To begin, we locate the 90% column and then run down this column to the twentieth row to find the number 1.725. According to this table, ninety percent of the area under the t distribution on twenty degrees of freedom lies between -1.725 and +1.725.

b. As a second illustration of the use of this table, let's find the value t such that 95% of the area under the t distribution on ten degrees of freedom lies between the value $-t$ and the value $+t$. To begin, we locate the 95% column and then run down this column to the tenth row to find the number 2.228. According to this table, ninety-five percent of the area under the t distribution on ten degrees of freedom lies between -2.228 and +2.228.

7.4 MEASUREMENT ERROR

7.4.1 THE NOTION OF MEASUREMENT ERROR

In Chapter 1, you had the opportunity to work in the presence of measurement error. In that instance, each student was asked to enumerate the number of occurrences of either the letter *e* or the letter *r* in a passage. Thus, a particular student may have counted ten e's too few, in which case the associated measurement error would have been -10. Another student may have counted five r's too many, in which case the associated measurement error would have been +5.

In this chapter, we will again investigate measurement error. In particular, each student will be asked to bisect each of twenty lines freehand. In other words, without the use of a ruler each student will be asked to estimate the midpoint of each of twenty lines. Each student will then be asked to measure the distance from the left endpoint of a given line to the estimated midpoint, for each of the twenty lines. The estimated distance will then be compared to the "true" distance, for each of the twenty lines.

If one were able to determine the exact midpoint of each of the 20 lines, there would be no measurement error in the determinations. But that is not possible, and one would expect measurement error in the determinations.

This example of measurement error is common to many scientific situations. In many scientific problems one presupposes the existence of a "true" value or physical constant (which we will refer to as μ). (One such enduring value, the velocity of light, will be discussed later in this section.) A basic scientific reality is that if one were to repeatedly measure this constant μ with a sufficiently sensitive apparatus, then one would find that the measurements change from determination to determination. Of course, it is believed that it is not the physical constant which changes from determination to determination, but rather it is the "noise" in the measurement process which gives rise to measurement errors. The goals in this section are to learn how to estimate μ and to learn how to prescribe a range of plausible values for μ, given a series of determinations of μ made in the presence of measurement error.

To motivate the development of the measurement error problem, let's formulate the measurement error problem in terms of drawing tickets from a hat. As an illustration, consider the line bisection experiment. In this experiment, consider a hat where written on each ticket in the hat is a *potential measurement error* to be incurred when a student attempts to bisect a line. Each time someone attempts to bisect a particular line, it is supposed that a ticket is selected at random from the hat and it is the value which is written on this ticket which is added to the *true distance* to give the *measured distance.*

More formally, suppose that for a particular line there is a true distance from the left endpoint to the midpoint of the line. (Let's call this value *d*.) Rather than exactly determining *d*, the recorded measurement is *D*, where

$$D = d + \varepsilon ,$$

where ε is the measurement error. Conceptually speaking, the quantity ε represents the "noise" which is introduced when one attempts to bisect a line. We are to think of this "noise" as a value which is randomly selected from a hat of "noise" values. It is this "noise" which contaminates a determination of *d*. Because the recorded measurement can be thought of as the sum of the enduring value and the error, the *variability in the measurements* serve as a guide to the *variability in the measurement errors*.

It is assumed that the measurement process is *unbiased* in the sense that while some errors are positive (selecting such an error would result in an *overassessment of d)* and some errors are negative (selecting such an error would result in an *underassessment of d), the average error is zero*. In other words, if one were to somehow average the values on the infinite number of error tickets in the hat, then one would find the answer to be 0.0. By assuming that the average measurement error is zero, it is believed that the measurement process yields measurements which are not *systematically* off the mark, that is, the determinations are not systematically too low or too high.

7.4.2 THE SPEED OF LIGHT

As an illustration of the concept of measurement error, consider an early attempt to measure the velocity of light. While it had been shown by Römer in 1676 that light travels at a finite velocity, it was not until the end of the nineteenth century that accurate determinations of the velocity of light were attempted. Over the period June 5, 1879 to July 2, 1879, A. A. Michelson made one hundred determinations of the velocity of light in air (Michelson, 1882).

Michelson's first measurement of the velocity of light, 299850 km/sec, can be viewed as the sum of the true velocity of light in air (μ) plus a measurement error (ε). Mathematically speaking,

$$1^{st}\,\textit{determination} = \text{true velocity} + \textit{error} \text{ or}$$
$$u_1 = \mu + \varepsilon_1$$

$$2^{nd}\,\textit{determination} = \text{true velocity} + \textit{error} \text{ or}$$
$$u_2 = \mu + \varepsilon_2$$

.

.

$$100^{th}\,\textit{determination} = \text{true velocity} + \textit{error} \text{ or}$$
$$u_{100} = \mu + \varepsilon_{100}$$

If one were to then average these one hundred determinations of the velocity of light in air (we will refer to this average as $\bar{u}$), then $\bar{u}$ would equal

$$\bar{u} = \mu + \bar{\varepsilon},$$

where $\bar{\varepsilon}$ equals the average of the one hundred ε's. In other words, if one averages the one hundred determinations, then $\bar{u}$ would differ from the true velocity μ by the average of the *selected* ε's. Since it is presumed that the average of all the values written on the tickets in the measurement error hat is 0.0, then the average of the measurement errors in the sample of size one hundred should be about 0.0.

How far would one expect the average of the measurement errors in the sample, $\bar{\varepsilon}$, to depart from 0.0? Clearly, the answer to this question depends on the variability of the values written on the tickets in the measurement error hat and the sample size. In particular, given a hat where the average of the values on the tickets in the hat is 0.0 and the *modified* standard deviation of the errors on the tickets in the hat is σ, we know that in repeated samples of size n the "typical" sample average of n errors will be about $\frac{\sigma}{\sqrt{n}}$ units from 0.0. In this way, given the results of *one sample,* one can expect $\bar{\varepsilon}$ to be about $\frac{\sigma}{\sqrt{n}}$ units from 0.0. Thus, the standard error of $\bar{\varepsilon}$ is $\frac{\sigma}{\sqrt{n}}$. Now note that because the average of the n determinations, $\bar{u}$, is equal to the sum of μ and $\bar{\varepsilon}$, we expect $\bar{u}$ to be about $\frac{\sigma}{\sqrt{n}}$ units from μ.

Since one expects $\bar{u}$ to be about $\frac{\sigma}{\sqrt{n}}$ units from μ, a reasonable candidate for a range of plausible values for μ is given as $\bar{u} \pm \frac{\sigma}{\sqrt{n}}$. In fact, it turns out that when the sample size n is sufficiently large, one can be 68% confident that the interval given by $\bar{u}$ plus-or-minus this standard error includes the value μ.

More generally, a 100α% confidence interval for μ is given by

$$(\bar{u} - z\frac{\sigma}{\sqrt{n}}, \bar{u} + z\frac{\sigma}{\sqrt{n}}),$$

where z is the value such that the area under the normal distribution from $-z$ to $+z$ is equal to 100α%, when the sample size is sufficiently large.

When σ is unknown, but the sample size is large, a 100α% confidence interval for μ would be given by

$$(\bar{u} - z\frac{SD}{\sqrt{n}}, \bar{u} + z\frac{SD}{\sqrt{n}}),$$

where SD is the standard deviation of the n determinations.

As an illustration of this methodology, the average of Michelson's one hundred observations is 299852.4 km/sec and the standard deviation is 79.01 km/sec. Thus, a range of plausible values for the speed of light in air, for which one is 95% confident includes the true velocity, is given by

$$(299852.4 - 1.96\frac{79.01}{\sqrt{100}}, 299852 + 1.96\frac{79.01}{\sqrt{100}}),$$

or (299836.91 km/sec , 299867.89 km/sec).

When the sample size n is not *sufficiently large,* and σ is not known, then one can use the t distribution to construct a $100\alpha\%$ confidence interval for μ, *assuming that the values on the tickets in the measurement error hat follow the normal distribution.* In other words, in such a situation one can use as a $100\alpha\%$ confidence interval for μ

$$(\bar{x} - t\frac{SD}{\sqrt{n}}, \bar{x} + t\frac{SD}{\sqrt{n}})$$

where t is the value such that the area under the t distribution on $n-1$ degrees of freedom from $-t$ to $+t$ is $100\alpha\%$.

7.4.3 THE EFFECT OF BIAS

Unfortunately, in real life we must recognize the fact that inherent in the measurement process is a certain amount of *bias*. As a child, I remember the corner fruit peddler who left a nickel on the scale when weighing my bag of grapes. In such a situation, rather than recording the weight of the bag of grapes plus the measurement error: $W = w + \varepsilon$, the peddler recorded $W = w + b + \varepsilon$, where w is the true weight of the bag of grapes, b is the weight of the the nickel (i.e. the bias) and ε is a measurement error. In other words, the peddler's recorded weights (W) *differ systematically* from the true weight (w) by the weight of the nickel (b). Given that the nickel is on the scale, there is no way that one can manipulate the *recorded* measurements $W_1, W_2, \ldots W_n$ to estimate the weight of the bag of grapes (w). In other words, given the series of recorded measurements one can compute the average of these values. However, this average would be an estimate of the weight of the bag of grapes *plus* the weight of the nickel.

Let's return to an examination of Michelson's data. The conclusion based upon that data is that we are 95% confident that the range of values specified by the interval (299836.91 km/sec, 299867.89 km/sec) includes the true velocity of light in air. But, modern instruments have determined the true velocity to be 299734.5 km/sec-- which does not fall in our 95% confidence interval. What happened was that the quantity Michelson measured was not the quantity he wanted to measure. That is, somewhere in his measurement process there was a bias that added, on average, about (299852.4 km/sec - 299734.5 km/sec =) +117.9 km/sec to each of his determinations.

The goal of this chapter's investigation will be a study of the *measurement error* which arises when one attempts to bisect a line. You will assess whether there is any systematic bias in this process, as well as determine whether the amount of bias varies from person-to-person. In the investigation, the true distance from the left endpoint of each line to the midpoint of that line will be known. Since for each line you will know d, you can subtract this value from the recorded measurement and examine what is "left over". You will examine what is "left over" to quantify the magnitude of the bias and to assess whether the bias varies with students or some other aspect of the experiment.

7.5 MATERIALS

Each student should have a ruler, a set of lines, and the data sheet.

7.6 EXPERIMENTAL PROTOCOL

1. On the following page is a set of twenty lines of unequal lengths. For each of these lines, mark the point which bisects the line. The bisection is to be done *freehand.*

2. After you have bisected all of the lines, measure the distance from the *left end of the line to the point which you marked* for each line. Use the ruler on the DATA SHEET page to perform this measurement.

3. Record these distances on the DATA SHEET.

Line 1: ______________________

Line 2: ____

Line 3: ____________________________________

Line 4: _______________________________

Line 5: _________

Line 6: ______________

Line 7: ____

Line 8: ___________________________

Line 9: _________________

Line 10: ______________________________

Line 11: _____

Line 12: ____

Line 13: ____________________

Line 14: ____

Line 15: _________________________

Line 16: ____

Line 17: _____________________________

Line 18: ____

Line 19: ____

Line 20: ___________

DATA SHEET NAME____________________

Line 1: ____________	Line 11: ____________
Line 2: ____________	Line 12: ____________
Line 3: ____________	Line 13: ____________
Line 4: ____________	Line 14: ____________
Line 5: ____________	Line 15: ____________
Line 6: ____________	Line 16: ____________
Line 7: ____________	Line 17: ____________
Line 8: ____________	Line 18: ____________
Line 9: ____________	Line 19: ____________
Line 10: ____________	Line 20: ____________

Hand in this page so that the data may be pooled.

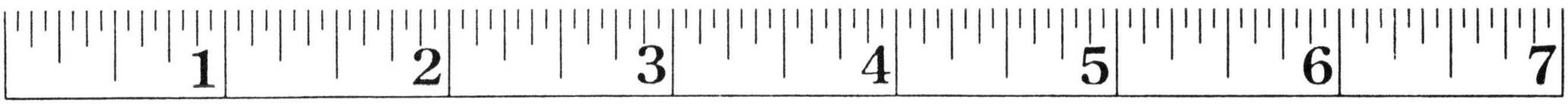

7.7 ANALYSIS

Your instructor will provide you with the class results, so that you may answer the following questions. In particular, you will be given a rectangular array of numbers with n rows corresponding to the n students who participated in the experiment and twenty columns corresponding to the twenty lines which each participant was to bisect. Each entry in the table represents the difference between a *measurement* and a *true* quantity. The *measurement* is the measured distance between the left endpoint and the *estimated* midpoint of a given line made by a particular student. The *true* quantity is the true half-length of that line. In this way, an entry in this table represents the *measurement error* for a particular student attempting to bisect a given line.

In question 7.1, you will examine the *measurements* from the point of view of cleaning and scrubbing the data. In other words, you will be asked to scan the data for observations which appear to have not been determined according to the protocol of the experiment. In questions 7.2 through 7.6, you will examine the bias in the class' measurement errors. In questions 7.7 through 7.9, you will examine the bias on a student-by-student basis. In questions 7.10-7.13, you will examine whether the bias in the measurement errors varies with the line length or the run order.

7.1. For each of the twenty lines, draw a histogram of the *n measurement errors*. Describe what you see - remember Shape, Location and Spread. For the outlying measurements, discuss possible explanations for the discrepancies. Should such points be omitted from further analysis or should they be adjusted in light of why they may be discrepant? How might the *goal of the analysis* impact on how the outliers should be treated? Are there any particularly error-prone lines?

7.2. Draw a histogram of all $20n$ measurement errors. Describe what you see. (Remember Shape, Location and Spread!) Is the histogram symmetric or somewhat skewed? Is there a point about which the histogram seems to be centered? Are there any outliers in the data? Relate these outliers to the outliers of 7.1.

7.3. Compute the average and median of all $20n$ measurement errors. Are they very different? Would you have expected this result, given your response in 7.2?

7.4. Of what quantity is your $\bar{x}$ an estimate? In other words, describe what you mean by the "population average" (μ) in this situation. (Hint: $\bar{x}$ and μ have to do with the *bias* incurred when attempting to bisect a line.)

7.5. Compute a 95% confidence interval for the average measurement error (based on all $20n$ data points). Should you use the normal or the t table to define your confidence number? Does it matter which table you use in this case? Interpret your 95% confidence interval in the context of this experiment.

7.6. Does your 95% confidence interval, that is, your range of plausible values for μ include the value 0.0? If yes, what does that imply about the bias in the measurement errors? If no, what does that imply about the bias in the measurement errors?

7.7. For each of the n students, compute the average of the twenty measurement errors. Draw a histogram of these n averages. Describe what you see. (Remember Shape, Location and Spread!) Is the histogram symmetric or is it somewhat skewed? About which point does the histogram seem to be centered? (Compute the average and median of these n averages.) Are there any outliers?

7.8. Compare the histogram of 7.7 with the histogram you drew in 7.2. Which histogram would you expect to be more tightly concentrated? Why? Which histogram would you expect to be more symmetric? Why?

7.9. For each of the n students, compute the 95% confidence interval based on their twenty measurement errors. Should you use the normal table or the t table to select your confidence number? What proportion of these confidence intervals include the value 0.0 as a plausible value? What proportion of the confidence intervals would you expect to cover 0.0 if there is no bias in the measurements? What can you conclude from these observations?

7.10. For the first line compute the average measurement error (over the n students who participated in the experiment). Repeat this for lines #2 through #20. Plot the average measurement error for each line versus the corresponding line number, that is, the measurement error for line #1 should be plotted against the number 1.0, the measurement error for line #2 should be plotted against the number 2.0, and so on. Do you see any trends in the measurement error as a function of the *order* in which the lines were bisected? Discuss.

7.11. Repeat question 7.10 using the median measurement error for each line. Compare these results with those results obtained using the average.

7.12. Plot the average measurement error for each line as a function of the *true* length of the line. (Your instructor will provide you with the true line lengths.) Do you see any trends in the measurement error as a function of the line *length?*

7.13. Repeat question 7.12 using the median measurement error for each line. Compare these results with those results obtained using the average. Discuss.

7.8 MINITAB APPENDIX

Given a rectangular array of numbers, with n rows corresponding to the n students who participated in the experiment and twenty columns corresponding to the twenty lines which each participant was to bisect, you will be asked to compute the average and the standard deviation of each participant's twenty measurement errors. To do this, you can use the RMEAN and RSTDEV commands in MINITAB. In particular, given that the data are stored in columns C1-C20, the command

> RMEAN of columns C1-C20 and put the row-wise averages in column C21

or more simply RMEA C1-C20 C21 , will compute the average of each row and will put these values in column C21. Moreover, the command

> RSTDEV of columns C1-C20 and put the row-wise standard deviations in column C22

or more simply RSTD C1-C20 C22 , will compute the standard deviation of each row and will put these values in column C22.

Given that column C1 contains data for which we wish to compute a 95% confidence interval (using the normal distribution to supply our confidence number) and given that σ is equal to 1.2, the command

> ZINTERVAL with 95 percent confidence taking the population standard deviation equal to 1.2 for the data in column C1

or more simply ZINT 95 1.2 C1 , will compute the confidence interval of interest.

Given that column C1 contains data for which we wish to compute a 95% confidence interval (using the *t* distribution to supply our confidence number), the command

> TINTERVAL with 95 percent confidence for the data in column C1

or more simply TINT 95 C1 , will compute the confidence interval of interest.

7.9 REFERENCE

Michelson, A. A. "Experimental Determination of the Velocity of Light Made at the United States Naval Academy, Annapolis." *Astronomical Papers, 1,* 109-145, 1882.

8

On the Face of It

Key Words: Hypothesis Testing; Null Hypothesis; Alternative Hypothesis; Reference Distribution; p-value; Test Statistic; Significance Level; t-test.

With seasonally adjusted temperatures, you could eliminate winter in Canada.
-Robert L. Stansfield

8.1 INTRODUCTION

Common to the design of the Great Pyramid of Cheops in Egypt, the Parthenon in Greece, a peculiar continued fraction, as well as the human form is the curious number $\frac{1+\sqrt{5}}{2} = 1.618\ldots$, which will be denoted by ϕ. The Great Pyramid of Cheops was built according to right triangular forms where the ratio of the hypotenuse of the triangle to the base of the triangle is ϕ. The facade of the Greek Parthenon was constructed so that the ratio of the height of the structure to the length of the front columns is ϕ. A mathematician can show that

$$\phi = 1 + \cfrac{1}{1 + \cfrac{1}{1 + \cfrac{1}{1 + \cdots}}} .$$

Fra Luca Pacioli in his book, *Divina Proportione* (which was illustrated by Leonardo da Vinci) outlines thirteen geometrical properties of ϕ. (One of the more mystical properties is that two crossing diagonals of a regular pentagon divide one another in the ratio $\phi : 1$.) Finally, it is said that for a "typical" person, if one measures the distance from the hairline to the chin and the distance from the right-most end of the right eyebrow to the left-most end of the left eyebrow, then the ratio of the larger to the smaller distance is ϕ. What we wish to explore is whether the human face follows the magic ϕ. In the following section, ϕ will be discussed in more detail. Following this presentation, some basic notions regarding hypothesis testing will be discussed.

8.2 THE GOLDEN PROPORTION

Where is the best point to cut a line segment? Pythagorus would probably cut the line segment AC at the point B.

The Best Place to Chop the Line AC
is at the Point B

A B C

According to the ancient geometers, the most aesthetically pleasing place to divide this line segment is the point *B*. It so happens that the ratio of the length of the entire line segment [AC] to the length of the longer segment [AB] is equal to the ratio of the length of the longer line segment [AB] to the length of the shorter segment [BC]. (As a general comment, we will denote the length of a line segment using brackets [ ].) In formula terminology, the point B cuts the line segment AC so that:

$$\frac{[AC]}{[AB]} = \frac{[AB]}{[BC]}.$$

It is for this reason that the point *B* forms the *golden proportion* of the line *AC*.

Note that if the length [BC] is one unit, then the length [AB] equals ϕ and the ratios $\frac{[AC]}{[AB]}$ and $\frac{[AB]}{[BC]}$, thereby, equal ϕ. Note that if we let [BC] = 1 and [AB] = x, then the golden proportion formula gives:

$$\frac{[AC]}{[AB]} = \frac{[AB]}{[BC]} \text{ which implies that } \frac{x+1}{x} = \frac{x}{1}.$$

The equation $\frac{x+1}{x} = \frac{x}{1}$ implies that $x^2 - x - 1 = 0$. If one were to then solve this quadratic equation for x, one would find that

$$x = \frac{1 \pm \sqrt{5}}{2}.$$

Now the root $x = \frac{1 - \sqrt{5}}{2}$ equals -0.618... , which is unacceptable as a length for the line segment [AB]. Thus, it is the root $x = \frac{1 + \sqrt{5}}{2} = 1.618...$ which is an acceptable length and it is this value which we have come to know as ϕ.

In the previous paragraphs, it was discussed where the ancients would cut a line. If one were to draw a rectangle of width w and length l, how should w and l be selected? As an answer to this question, Pythagorus would select w and l such that

$$\frac{l}{w} = \frac{l+w}{l} \quad .$$

If one were to manipulate this equation, as was done to the golden proportion formula in the preceeding paragraph, it could be shown mathematically that this equation implies that

$$\frac{l}{w} = \frac{1 + \sqrt{5}}{2} = 1.618... = \phi \ .$$

In other words, if l and w satisfy

$$\frac{l}{w} = \frac{l+w}{l} \quad ,$$

then it mathematically follows that l and w are in a ratio of ϕ to 1. Again, the magic ϕ appears. Rectangles which are constructed such that $\frac{l}{w} = \phi$ are called *golden rectangles*. One could continue describing all the peculiar situations where the value ϕ arises - but we will not. If this discussion piqued your interest in ϕ, then you may wish to read Chapter 8 of Gardner (1961). This chapter's investigation will focus on the issue of whether the rectangle which encloses the human face (forehead to chin - eyebrow to eyebrow) follows the golden rectangle rule.

8.3 THE HYPOTHESIS TEST

8.3.1 THE NULL HYPOTHESIS

In hypothesis testing, one begins with the null hypothesis. It is the **null hypothesis** which makes some statement about the values in the population. To illustrate the concept of the null hypothesis in the context of the population average, consider as the population of interest, the population of picture frames at the Elvehjem Museum of Art in Madison, Wisconsin. What we wish to assess is whether these frames are constructed so that the length-to-width ratio are in accord with ϕ.

Consider a hat containing tickets. Written on each of these tickets is the quantity $\frac{l}{w}$, corresponding to the exact length of a particular picture frame divided by the exact width of the picture frame. (Each frame in the museum is represented by one ticket in the hat.) One may ask whether the average (μ) of all the quantities in the hat (the average of the $\frac{l}{w}$'s over the population of all frames) is equal to ϕ. In other words, if one were to consider the population of all picture frames at the Elvehjem Museum, one may wonder whether the "typical" frame is constructed according to the golden rectangle rule. Thus, the *null hypothesis* for this problem is that the average (μ) of all the quantities $\frac{l}{w}$ in the hat is equal to ϕ and any departure from ϕ in the sample is due to chance. One can express this sentence in formula notation as:

$$H_o: \mu = \phi \, .$$

Why is this hypothesis called the *null* hypothesis? This hypothesis indicates that there is no *real* effect in the data - it is only due to *chance* that the data differ from what is predicted by the null hypothesis.

8.3.2 THE ALTERNATIVE HYPOTHESIS

In fact, the null hypothesis that the average length-to-width ratio is equal to 1.618 *may be false*. In other words, the discrepancy between the data and what the null hypothesis predicts may be due to more than just chance fluctuation. In particular, one of the following alternative hypotheses may describe the "true" state of affairs:

a. The average length-to-width ratio (μ) is *less* than 1.618.

b. The average length-to-width ratio (μ) is *greater* than 1.618.

c. The average length-to-width ratio (μ) is *either less than* or *greater than* 1.618.

Statisticians have given a name to this second set (a,b, or c) of possible explanations of the true state of nature. Statisticians refer to any of these alternative hypotheses that the "typical" length-to-width ratio for the population of picture frames at the Elvehjem Museum is some value other than 1.618 as an *alternative hypothesis* to the null hypothesis. Thus, the null hypothesis specifies what one is to believe if the data set is no different from what the "theory" predicts. The **alternative hypothesis** specifies what one is to believe if, in fact, the null hypothesis is false.

Mathematically speaking, the alternative hypothesis *a* corresponds to

$$H_a: \mu < 1.618,$$

the alternative hypothesis b corresponds to

$$H_a\colon \mu > 1.618,$$

while the alternative hypothesis c corresponds to

$$H_a\colon \mu \neq 1.618.$$

Statisticians refer to alternative hypotheses a and b as *one-sided alternative hypotheses,* while alternative hypothesis c is a *two-sided alternative hypothesis.* What will be discussed in the following section is how one is to assess whether the null hypothesis or one of these alternative hypotheses is more congruent with the data.

8.3.3 THE BASIC PHILOSOPHY

Having specified the null hypothesis and the alternative hypothesis, one has to assess which of these suppositions is most congruent with the data. The approach that is adopted in statistics is to assume the null hypothesis to be true. One then compares what the null hypothesis predicts with what the data indicate. If the two are in discord, then one can reject the null hypothesis in favor of an alternative hypothesis.

> In the **hypothesis test,** one examines whether the discrepancy between the data and the null hypothesis can be attributed to chance.

8.3.4 THE TEST STATISTIC

Assume one draws a sample of size 100 from a hat where σ is equal to 2.0 and finds that the average width-to-length ratio in the sample is 2.418. How is one to measure the discrepancy between the data and the null hypothesis? Assuming the null hypothesis correctly describes the average of the tickets in the hat, the data are

$$\frac{2.418 - 1.618}{2.0/\sqrt{100}} = 4.0$$

standard errors from the null hypothesis.

In statistics there is a formal approach for assessing the extent to which the data and the null hypothesis are in discord. This yardstick is called a test statistic.

> The **test statistic** measures the distance between the data and the null hypothesis, assuming the null hypothesis to be correct.

In the picture frame example, the test statistic tells us that the data are four standard errors from the null hypothesis.

8.3.5 THE *p-value*

Once one has defined a test statistic for measuring this distance, one must determine whether a particular distance is large. Of course, since one is dealing with the results of experiments which vary from repetition to repetition of the same experiment, one cannot focus directly on the magnitude of this distance, but rather on the chance of observing a more extreme result. That is, one must have a way of assessing the probability of observing a distance (between the data and the null hypothesis) as extreme or more extreme than what was actually observed. It is this probability which is called the *significance level* or *p-value* of the hypothesis test.

> The **p-value** is the probability of observing an outcome as extreme or more extreme than what was actually observed, assuming the null hypothesis to be correct.

In terms of the picture frame example, note that the area under the normal distribution to the right of 4.0 is quite small (about six parts in one hundred thousand). Thus, it seems very unlikely that the tickets in the sample were drawn from a hat where the average is 1.618.

8.3.6 DRAWING A CONCLUSION

In performing a hypothesis test one must specify a means of measuring the distance between how the data came out and how the null hypothesis implies that the data should have come out. In other words, one must specify a test statistic. Having defined the test statistic, one computes the statistical significance of the test, that is, the probability of observing a value of the test statistic which is more extreme than what was observed, assuming the null hypothesis to be true.

> The smaller the *significance level or p-value,* the more evidence we have *against* the null hypothesis.

Why does a small *p-value* provide evidence against the null hypothesis? The idea behind hypothesis testing is that of proof by contradiction - *reductio ad absurdum.* In hypothesis testing, one assumes the null hypothesis to be true. Based on this assumption, one computes the probability that the difference between the data and the null hypothesis is due to chance fluctuation. If this probability is small, then it is unlikely that the difference between the data and the null hypothesis can be explained by chance fluctuation. In such a context, something else besides chance is contributing to the observed discrepancy between the data and what the null hypothesis is predicting. On the other hand, if the *p-value* is not small, then one has little evidence to suggest that something other than chance is behind the observed discrepancy between the data and what is predicted by the null hypothesis.

8.3.7 THE TESTING PROCESS

To illustrate the ideas of null hypothesis, alternative hypothesis, test statistic, and p-value, let's test the null hypothesis that the typical length-to-width ratio of the population of picture frames is equal to 1.618.

To implement the hypothesis test, one must proceed through the following six steps:

1. Specify H_o.
2. Specify H_a.
3. Do the experiment!
4. Compute the value of the test statistic.
5. Determine the appropriate reference distribution.
6. Compute the *p-value*.

What is H_o? The "theory" predicts that the average of the length-to-width ratios of the picture frames at the Elvehjem Museum of Art should equal 1.618, so the null hypothesis is H_o: $\mu = 1.618$. What is H_a? Assume that one had some *a priori* reason to suspect that if the null hypothesis is wrong, then the "typical" length-to-width ratio of the picture frames in the population is some value less than 1.618. Thus, the alternative hypothesis is H_a: $\mu < 1.618$. What is the value of the test statistic? Assume that a sample of size 100 is drawn from the hat of length-to-width ratios in which it is known that $\sigma = 5.0$ and that the average length-to-width ratio in the sample is 1.418. Plugging these numbers into the test statistic equation, we have that

$$\frac{\bar{x}-\mu_o}{\sigma/\sqrt{n}} = \frac{1.418-1.618}{5.0/\sqrt{100}} = -1.0.$$

Is this value of the test statistic -1.0 statistically significant? In other words, what is the probability of observing a result as extreme or more *extreme* than -1.0? To compute this probability, one must define what is meant by the word extreme. The definition of the word extreme depends on how one *a priori* specified the alternative hypothesis. By specifying the alternative hypothesis as H_a: $\mu < 1.618$, one looks for evidence against the null hypothesis in the direction of large negative values of the test statistic. As such, for this alternative hypothesis, the *p-value* is given as the area under the normal distribution to the left of -1.0. An examination of the normal table reveals that this probability is about 0.16. Based on this calculation, there is little evidence to reject the null hypothesis in favor of this alternative hypothesis.

In this example, it is assumed that σ is known. In the case where σ is unknown, but n is larger than 40, one may compare

$$\frac{\bar{x}-\mu_o}{SD/\sqrt{n}}$$

to the normal distribution to compute the *p-value*.

If one had no idea what σ should be and n is not *sufficiently large,* then it turns out that one can no longer use the normal distribution as the reference distribution. The correct reference distribution for the test statistic

$$\frac{\bar{x}-\mu_o}{SD/\sqrt{n}}$$

in such a case is a **t distribution.** More specifically, if σ is not known, but *the distribution of the values in the population does follow the normal distribution,* then one may use the t distribution on d degrees of freedom to assess the statistical significance of the result.

How many degrees of freedom should we use? If the sample size is n, then the correct t distribution to be used as the reference distribution for this test statistic is the t distribution with $d = n-1$ degrees of freedom.

To illustrate how to use the t table to compute *p-values,* consider the following examples:

a. Assume that one obtains a test statistic, based on 11 observations, equal to 1.812. For the alternative hypothesis H_a: $\mu > \mu_o$, one is interested in computing the probability of observing a larger value of this test statistic. Assuming that the data in the population follow the normal distribution, the correct reference distribution for this test statistic is the t distribution on 10 degrees of freedom. The t table tells us that the area on the t distribution on 10 degrees of freedom, between -1.812 and 1.812 is equal to 0.90. Thus, the area to the right of 1.812, under the t distribution on 10 degrees of freedom is equal to 0.05.

b. Assume that the test statistic is equal to -1.812. For the alternative hypothesis H_a: $\mu < \mu_o$, one is interested in computing the probability of a more negative value of the test statistic, i.e. the area under the t distribution on 10 degrees of freedom to the *left* of -1.812. Since all the t distributions are symmetric, the area under the t distribution to the left of -1.812 is equal to the area under the to the right of 1.812 or 0.05.

c. Assume that the test statistic, based on 11 observations, is equal to 1.95. For the alternative hypothesis H_a: $\mu > \mu_o$, one is interested in computing the area under the t distribution on 10 degrees of freedom to the right of 1.95. As was pointed out above, the area under the t distribution on 10 degrees of freedom between -1.812 and 1.812 is equal to 0.90. Also note that the area under this distribution between -2.228 and 2.228 is 0.95. In this way, the area to the right of 1.812 is equal to 0.05 and the area to the right of 2.228 is equal to 0.025 Since the t table does not have an entry of 1.95 for 10 degrees of freedom, one can only comment that the area to the right of 1.95 for the t distribution on 10 degrees of freedom is somewhere between 5% and 2.5%.

8.4 MATERIALS

Each student should have a ruler and the data sheet.

8.5 EXPERIMENTAL PROTOCOL

In this experiment, you will test the null hypothesis that the rectangle which encloses the human face follows the magic ϕ. In particular, the distance from the hairline to the tip of the chin (we will call this value l) will be measured, as well as the distance from the right end of the right eyebrow to the left end of the left eyebrow (we will call this value w). In the presence of measurement error, as well as in the presence of person-to-person variability in facial structure, you will attempt to assess whether the human face follows ϕ.

If you wear glasses, please remove your glasses before your hairline-to-chin and eyebrow-to-eyebrow distances are measured.

For bearded individuals, measure the distance to the tip of the beard. For balding individuals, measure the distance to the *present hairline*. For individuals who have colored their eyebrows, use the eyebrow defined by the coloring.

When measuring the hairline-to-chin or the eyebrow-to-eyebrow distances, DO NOT BEND THE RULER to the contour of the face. You are to measure the distances as they would be projected onto a plane tangent to the tip of the nose.

DATA SHEET NAME________________________

1. Each student should measure his or her own hairline-to-chin and eyebrow-to-eyebrow distances. Record the result.

__________ l **(hairline-to-chin)** __________ w **(eyebrow-to-eyebrow)**

2. Ask the person to your right to measure your hairline-to-chin and eyebrow-to-eyebrow distances. Don't let this person see your results until their measurements have been recorded. (If there is nobody to your right, then ask your nearest neighbor.) Record the result.

__________ l **(hairline-to-chin)** __________ w **(eyebrow-to-eyebrow)**

3. Record your sex.

__________ **male** __________ **female**

4. Do you have a beard?

__________ **yes** __________ **no**

5. Are you balding?

__________ **yes** __________ **no**

Hand in this page so the data may be pooled.

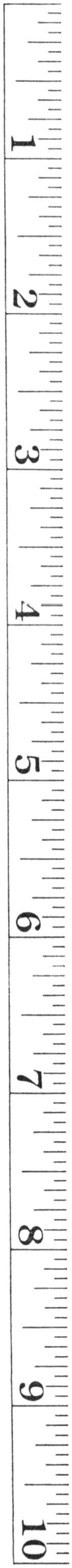
1
2
3
4
5
6
7
8
9
10

8.6 ANALYSIS

Your instructor will provide your with the class results, so that you may answer the following questions. In particular, you will be a rectangular array of numbers, corresponding to the questions in Section 8.5. In particular, if n students participated in the study, then the array will have n rows and seven columns. In questions 8.1-8.5, you are asked to examine how the data distribute. Problems 8.6-8.9 focus on the hypothesis test, while in problems 8.10-8.11 you are asked to investigate whether there is a sex effect.

8.1. Compute the n values of $\frac{l}{w}$ based on the *self-measured* values of l and w. Draw a histogram or stem-and-leaf plot of these n values. What do you see? (Remember Shape, Location and Spread!) Is the distribution of the values *skewed?* Are there any *outlying values* (i.e. values which are somewhat removed from the bulk of the data) in the data set? Can the outlying values be explained on the basis of the beard information? Can the outlying values be explained on the basis of the balding information? Do you think it justifiable to drop such outlying values from any further analysis?

8.2. Repeat question 8.1 using the *partner-measured* $\frac{l}{w}$ values.

8.3. Based on your plot in 8.1, is it necessary to transform the *self-measured* data so that the distribution of the observations is more symmetric? Based on your plot in 8.2, is it necessary to transform the *partner-measured* data so that the distribution of the observations is more symmetric?

8.4. If it is necessary to transform the *self-measured* $\frac{l}{w}$ values, then try the inverse $(\frac{1}{x})$, the logarithm $(\log_e(x))$, and the square root $(\sqrt{x})$ transformations. For the inverse transformation, draw the histogram of the transformed data. Now draw the analogous plots for the logarithm and square root transformations. Do any of the transformations seem to symmetrize the distribution of the data set? On which scale, would you suggest that the data be analyzed?

8.5. Repeat question 8.4 for the *partner-measured* $\frac{l}{w}$ values.

8.6. Test the null hypothesis that the "typical" human face follows ϕ using the *self-measured* $\frac{l}{w}$ values. What is your alternative hypothesis? What is the value of your test statistic? What is your *p-value?* What do you conclude? Transform the data using the transformation which you selected in problem 8.4. Repeat the hypothesis test on this scale.

8.7. Repeat question 8.6 using the *partner-measured* data. What do you conclude?

8.8 Does your conclusion using the *partner-measured* data differ from the conclusion you reached using the *self-measured* data? In other words, did you reject the null hypothesis in one case and not the other? If you answer in the affirmative, why do you suppose you reached different conclusions? (Compute the standard deviation of the $\frac{l}{w}$ values for the *self-* and *partner-* measured data sets and compare these two values. Would you expect one of these two values to be much larger than the other?)

8.9 In questions 8.6 and 8.7, what did you assume regarding *how the data were collected* to justify the interpretation of the *p-value?* Is this assumption reasonable?

8.10 Test whether the "typical" *male* face follows ϕ using both the *self-measured* and the *partner-measured* data sets. Be sure to check for any outlying values, as well as skewness in the distribution of the male $\frac{l}{w}$ values. If there are any outlying values, can the beard or balding information explain why these values are outlying? If there is a noticeable skewness to the distribution of the data, try to normalize the data using one of the various transformations presented in problem 8.4. Analyze the data on the more appropriate scale.

8.11 Repeat question 8.10 for the women. Do you obtain different results for the men than for the women? In other words, do you reject the null hypothesis for the men and not for the women or visa versa? If you answer in the affirmative, can you explain what aspects of the data led you to these divergent conclusions?

8.7 MINITAB APPENDIX

Given that column C1 contains data for which we wish to test the hypothesis that H_o: $\mu = 613.613$ (using the normal distribution as our reference distribution) and given that σ is equal to 1.2, the command

> ZTEST of 613.613 taking the standard error equal to 1.2 for the data in column C1

or more simply ZTEST 613.613 1.2 C1 , will perform the hypothesis test of interest and return the *p-value* for the alternative that H_a: $\mu \neq 613.613$.

Given that column C1 contains data for which we wish to test the hypothesis that H_o: $\mu = 613.613$ (using a *t distribution* as our reference distribution), the command

TTEST of 613.613 for the data in column C1

or more simply TTEST 613.613 C1 , will perform the hypothesis test of interest and return the *p-value* for the alternative that H_a: $\mu \neq 613.613$.

To compute the area to the *left of a value V* under the *t* distribution on *M* degrees of freedom, use the command to compute the

CDF of *V*;
T distribution on *M* degrees of freedom.

or more simply

CDF *V*;
T *M*.

8.8 REFERENCES

1. Gardner, M. *The 2nd Scientific American Book of Mathematical Puzzles and Diversions,* Simon and Schuster, New York, 1961.

2. Pacioli, L. *Divina Proportione.* Venice, 1509.

9

Aerobic Statistics

Key Words: Two-Sample Problem; Randomization Design; Population Design; Measurement Error Design; Confidence Interval; Standard Error of the Difference; Welch Two-Sample t Test; Pooled Two-Sample t Test; Hypothesis Test;

Another thing that cigarette smoking causes is statistics.
-Anonymous

9.1 INTRODUCTION

This chapter continues with a discussion of hypothesis testing and confidence intervals. The new twist is that the two-sample problem is considered. One goal is to determine whether there is a statistical difference between the two groups. Another goal is to specify a range of plausible values for the difference. The presentation begins with a discussion of various experimental designs for the two-sample problem. We will then introduce the appropriate reference distributions which will be used for testing, as well as for supplying confidence numbers.

To illustrate some of the ideas in this chapter, the students in your class will be randomized to one of two approaches for measuring their own pulse rate. Based on the results of the investigation, you will assess whether there is a statistical difference between these two techniques, define a range of plausible values for this difference, as well as explore other features of the data.

9.2 EXPERIMENTAL DESIGNS FOR THE TWO SAMPLE PROBLEM

There are three experimental designs for the two-sample problem: the population design, the randomization design and the measurement error design. The present goal is to discuss each of these experimental designs.

9.2.1 THE POPULATION DESIGN

In the population design, it is assumed that n_1 subjects are randomly selected from a particular population and n_2 subjects are randomly selected from another population. Because the units or subjects are randomly selected from the populations, the samples are representative of the respective populations. In terms of hats, consider two hats from which one randomly selects n_1 tickets from one hat and n_2 tickets from the other hat. Based on the averages of the values on the tickets in the two samples, namely $\bar{x}_1$ and $\bar{x}_2$, one wishes to assess whether the population averages (the averages of the tickets in the hats), μ_1 and μ_2, differ and to specify a range of plausible values for the difference $\mu_1 - \mu_2$. To this end, let σ_1 and σ_2 denote the respective population modified standard deviations.

9.2.2 THE RANDOMIZATION DESIGN

In this design, it is assumed that there are N subjects, of which n_1 are assigned to one treatment and n_2 are assigned to the other treatment, where $n_1 + n_2 = N$. One goal in this set-up is to determine whether there is a treatment effect. One wishes to determine whether the *observed* difference between the two groups is more extreme than that which is expected by *chance*. A second goal is to specify a range of plausible values for the magnitude of the treatment effect.

This randomization design was encountered in Chapter 1 when each member of your class was randomized to count e's or to count r's. To determine whether there is a treatment effect, you first computed the average measurement error for the *e* group ($\bar{x}_e$) and the average measurement error for the *r* group ($\bar{x}_r$). You then compared the *observed* difference ($\bar{x}_r - \bar{x}_e$) to the reference distribution of *potential* differences to assess the statistical significance of this observed difference. Given that relatively few of the potential differences were as extreme or more extreme than the observed difference, one would conclude that there is a treatment effect.

It is important to note that *chance* enters into the randomization design as a means of allocating the subjects or units under study to one of two treatments. Moreover, in the randomization design, the mechanism for the selection of the units or subjects into the study is not specified. In other words, in the randomization design one is given N subjects or units and one can make no assumptions as to how these subjects or units were obtained. Thus, there is no statistical basis in the randomization design to allow one to extrapolate to some population of interest. Any conclusions that are drawn from the data of a randomization design regarding the effectiveness of the treatment will refer only to the *specific units or subjects in the study*.

9.2.3 THE MEASUREMENT ERROR DESIGN

In Chapter 7, you performed an experiment in which you were to bisect a series of lines freehand. Note that no one was randomized in any way (that is, the design was not a randomization design), nor was anyone randomly selected from a population (that is, the design was not a population design). In that situation, *chance* entered via measurement errors. The hat contained an infinite number of tickets, each ticket labeled with a *potential* error. At your i^{th} attempt to bisect a line freehand, you conceptually drew a ticket from this hat and then recorded the sum of the true half length of the line *(d)* and this error (ε_i) as your guess (D_i) for the half-length of the line. Mathematically speaking, you recorded D_i, where

$$D_i = d + \varepsilon_i \ .$$

In the measurement error design, one considers two series of measurements based on either two methods of measurement of the same object or repeated measurement of two objects. In other words, one imagines two hats each with an infinite number of tickets. To measure using method #1 (or to measure unit #1) one conceptually draws a ticket from hat #1 and then records the sum of the true measurement (μ_1) and this error (ε_i) as the measurement (u_{1i}), *i.e.*

$$u_{1i} = \mu_1 + \varepsilon_i \ .$$

Analogously, to measure using method #2 (or to measure unit #2) one conceptually draws a ticket from hat #2 and then records the sum of the true measurement (μ_2) and this error (ε_i) as our measurement (u_{2i}), *i.e.*

$$u_{2i} = \mu_2 + \varepsilon_i' \ .$$

Based on a random selection of tickets from these two hats, that is, two series of measurements u_{1i} and u_{2i}, where

$$u_{1i} = \mu_1 + \varepsilon_i \ , \ i = 1, \ \cdots \ , n_1$$
$$u_{2i} = \mu_2 + \varepsilon_i \ , \ i = 1, \ \cdots \ , n_2 \ ,$$

one wishes to determine whether the measure of an item via method #1 (μ_1) differs from that for method #2 (μ_2) or whether the measure of item #1 (μ_1) differs from the measure of item #2 (μ_2). Another goal in this design is to specify a range of plausible values for the difference μ_1 - μ_2, based on the information in the two samples. To this end, let σ_1 and σ_2 denote the respective modified standard deviations of the error tickets in the two hats.

9.3 ASSESSING DIFFERENCES

We now proceed with a brief presentation of the statistical methodology for testing for differences and for defining a range of plausible values for the difference. Regarding testing, take as the null hypothesis

$$H_o\colon \mu_1 = \mu_2 \ ,$$

where μ_1 is the average of the values on the tickets in hat #1 - the population #1 average and μ_2 is the corresponding value for the second population. As in the one-sample context, to test this hypothesis one must define a measure of the distance between what was observed and how the null hypothesis predicts the data should look, i.e. a test statistic. We will then present the reference distribution for the test statistic, constructed under the assumption that the null hypothesis is true. This reference distribution will be used to assess whether the discrepancy between the data and the null hypothesis can be attributed to chance. If the probability of observing a more extreme value of the test statistic, the *p-value* is small, then it is unlikely that this observed discrepancy is due to chance. If this *p-value* is not small, then one has little evidence against the null hypothesis.

In addition to testing, one is interested in specifying a range of plausible values for the difference $\mu_1 - \mu_2$, based on the information in the two samples. As in the one-sample context, one begins with the *standard error.* It is this quantity which quantifies the + or - *something* which one can attach to the estimate of the difference. One can then define the appropriate *confidence number* which indicates *the degree of confidence* that the range of plausible values for the difference includes the value $\mu_1 - \mu_2$.

As in the one-sample case, it turns out that the selection of the test statistic, the standard error, and a reference distribution for the test statistic and for supplying our confidence number will depend on the number of observations *in the sample* and our knowledge of σ_1 (the modified standard deviation of the values in population #1) and σ_2 (the modified standard deviation of the values in population #2). As such, six cases are considered: ("Small sample" refers to the case when at least one of the two sample sizes is less than 40.)

CASE	SAMPLE SIZE	σ_1 AND σ_2
#1	*large sample*	$\sigma_1 = \sigma_2 = \sigma$ (σ known)
#2	*large sample*	$\sigma_1 = \sigma_2 = \sigma$ (σ unknown)
#3	*large sample*	$\sigma_1 \neq \sigma_2$ (σ_1 and σ_2 are known)
#4	*large sample*	$\sigma_1 \neq \sigma_2$ (σ_1 and σ_2 are unknown)
#5	*small sample*	$\sigma_1 = \sigma_2 = \sigma$ (σ unknown)
#6	*small sample*	$\sigma_1 \neq \sigma_2$ (σ_1 and σ_2 are unknown)

It turns out, that the confidence interval for each of these cases follows a generic form. The GENERIC CONFIDENCE INTERVAL FORMULA for $\mu_1 - \mu_2$ is given as

$$(\bar{x}_1 - \bar{x}_2 - CN{\cdot}\mathit{Standard\ Error}, \bar{x}_1 - \bar{x}_2 + CN{\cdot}\mathit{Standard\ Error}) .$$

In other words, to the estimate of the difference $\bar{x}_1 - \bar{x}_2$ one tacks on a *plus or minus something.* This *something* consists of the standard error or estimated standard error of $\bar{x}_1 - \bar{x}_2$ (which quantifies how well determined is $\mu_1 - \mu_2$ based on the data in the two samples) and a confidence number (which specifies the degree of confidence one has that $\mu_1 - \mu_2$ is contained in the interval). Thus, one should realize that while the specifics of the *standard error* and the *confidence number* varies among cases, *the general form of the interval remains constant.*

There is a corresponding parallel regarding the test statistic. All test statistics for the above six cases share the GENERIC TEST STATISTIC FORMULA

$$\frac{\bar{x}_1 - \bar{x}_2}{\mathit{Standard\ Error\ of\ the\ Difference}} .$$

In general to assess the distance between the data ($\bar{x}_1 - \bar{x}_2$) and the null hypothesis (H_o: $\mu_1 - \mu_2 = 0.0$), one divides the quantity $\bar{x}_1 - \bar{x}_2$ by its standard error or by its estimated standard error. If the resulting ratio is *large,* (both its sign and magnitude) then there is evidence that the data are somewhat further from the null hypothesis than can be expected by chance. What is meant by the *standard error* and what is meant by *large* depends on the context. While the specifics of hypothesis testing varies with the context, the generalities remain constant.

The following table presents the appropriate standard error or estimated standard error for each of the cases which we have considered. The appropriate test statistic/confidence interval is obtained by simply plugging in the appropriate standard error (or estimated standard error) into the GENERIC TEST STATISTIC / CONFIDENCE INTERVAL FORMULA.

CASE	SAMPLE SIZE	σ_1 AND σ_2	SE or ESE	REFERENCE CURVE
#1	*large sample*	$\sigma_1 = \sigma_2 = \sigma$ (σ known)	$\sigma\sqrt{\frac{1}{n_1}+\frac{1}{n_2}}$	*Normal*
#2	*large sample*	$\sigma_1 = \sigma_2 = \sigma$ (σ unknown)	$SD_p\sqrt{\frac{1}{n_1}+\frac{1}{n_2}}$	*Normal*
#3	*large sample*	$\sigma_1 \neq \sigma_2$ (σ_1 and σ_2 are known)	$\sqrt{\frac{\sigma_1^2}{n_1}+\frac{\sigma_2^2}{n_2}}$	*Normal*
#4	*large sample*	$\sigma_1 \neq \sigma_2$ (σ_1 and σ_2 are unknown)	$\sqrt{\frac{SD_1^2}{n_1}+\frac{SD_2^2}{n_2}}$	*Normal*
#5	*small sample*	$\sigma_1 = \sigma_2 = \sigma$ (σ unknown)	$SD_p\sqrt{\frac{1}{n_1}+\frac{1}{n_2}}$	*t on $n_1 + n_2 - 2$ Degrees of Freedom*
#6	*small sample*	$\sigma_1 \neq \sigma_2$ (σ_1 and σ_2 are unknown)	$\sqrt{\frac{SD_1^2}{n_1}+\frac{SD_2^2}{n_2}}$	*t – see below for the Degrees of Freedom Formula*

For CASE #2 and CASE #5, the quantity SD_p is defined as:

$$SD_p = \sqrt{\frac{(n_1-1)SD_1^2+(n_2-1)SD_2^2}{n_1+n_2-2}}.$$

One refers to the test for CASE #5 as the **pooled two-sample t-test,** while the test for CASE #6 is called the **Welch two-sample t-test.** Regarding CASE #5 and CASE #6, it turns out that *if the data in both hats follow the normal distribution,* then the reference distribution for this test statistic is the t distribution. For CASE #5, the appropriate degrees of freedom are $n_1 + n_2 - 2$. For CASE #6, the appropriate degrees of freedom are

$$\frac{\left(\frac{SD_1^2}{n_1} + \frac{SD_2^2}{n_2}\right)^2}{\frac{SD_1^4}{n_1^2(n_1 - 1)} + \frac{SD_2^4}{n_2^2(n_2 - 1)}}.$$

(This is usually not an integer. In such cases one uses the next largest integer as the degrees of freedom.)

As an example, suppose the data in each sample seem to follow the normal distribution, one of the two sample sizes is less than 40 (i.e. a small sample case) and the sample standard deviations differ by a factor in excess of two (i.e. one has evidence to suggest that $\sigma_1 \neq \sigma_2$). The appropriate test statistic in this case is

$$\frac{\bar{x}_1 - \bar{x}_2}{\sqrt{\frac{SD_1^2}{n_1} + \frac{SD_2^2}{n_2}}}.$$

To assess the statistical significance of the value of this test statistic, one compares it to the *t* distribution on the degrees of freedom listed for CASE #6. If the *p-value* is small, then either one observed a rare event or the assumption that $\mu_1 = \mu_2$ is wrong.

9.4 MATERIALS

Each student should have a data sheet and a coin (preferably new and of the same denomination for each student).

In this experiment, you will be required to measure your pulse. There are two common approaches for measuring pulse rate.

THE RADIAL METHOD

In the *RADIAL method,* you are to measure the pulse at your right wrist using the tips of your index and middle fingers of your left hand. If you are wearing a wrist watch or bracelet on you right hand, please remove it. Begin by placing your fingertips on your right wristbone at the base of the thumb. Now move your fingertips toward your wrist. You should now feel your pulse in the pocket of skin.

THE CAROTID METHOD

In the *CAROTID method,* you are to measure the pulse on the right side of your neck using the tips of your index and middle fingers of your right hand. In particular, place your fingertips on your right jaw below your right earlobe. Now move your fingertips toward your neck. You should now feel your pulse about half way between your jaw and your neck at the carotid artery.

Using the RADIAL method practice counting your pulse.

Using the CAROTID method practice counting your pulse.

RADIAL

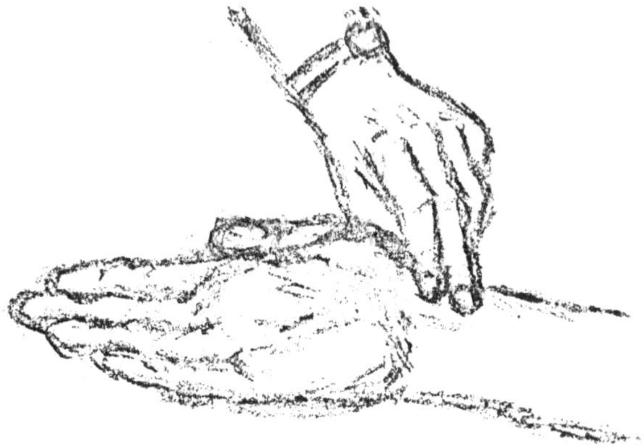

CAROTID

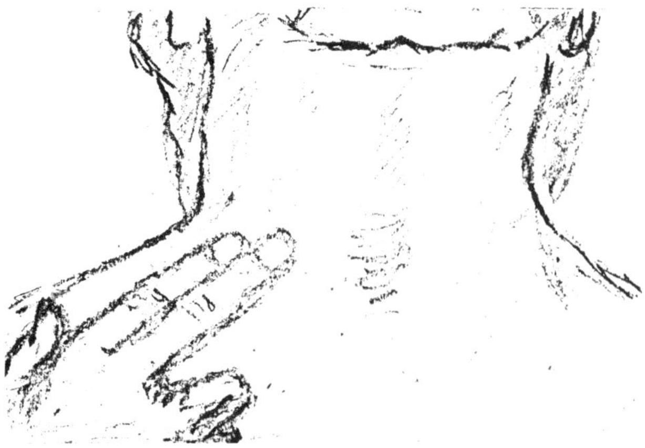

9.5 EXPERIMENTAL PROTOCOL NAME______________________

DATA SHEET

1. Record your height in **inches** ______________.

2. Record your weight in **pounds** ______________.

3. Record your GPA (A = 4.0) ______________.

4. Record your sex ______________.

   MALE = 1

   FEMALE = 2

5. Level of physical activity ______________.

   1 = works out three times
   or more per week

   2 = works out less frequently
   than
   three times per week

6. Smoking Habits ______________.

   1 = smokes about 2 packs or more per week

   2 = smokes less than 2 packs per week

7. Flip the coin. Record the result.

   ______ **HEAD** ______ **TAIL**

8. If the toss resulted in a **HEAD,** then you are to measure your pulse by the **RADIAL METHOD.**

If the toss resulted in a **TAIL,** then you are to measure your pulse by the **CAROTID METHOD.**

9. Using **your assigned method** for measuring your pulse, count the number of beats. Your instructor will signal when to start counting and when to stop counting.

I counted _______________ beats in thirty seconds.

10. Perform the physical activity. (The performance of this physical activity is voluntary. You are not required to perform the physical activity. IF YOU DO NOT perform the activity, then DO NOT answer the following question.)

11. Using **your assigned method** for measuring your pulse, count the number of beats. Your instructor will signal when to start counting and when to stop counting.

I counted _______________ beats in thirty seconds.

Hand in this page so that the class results may be pooled.

9.6 ANALYSIS

Your instructor will provide you with the class results, so that you may answer the following questions. You will be given a rectangular array of data, with n rows corresponding to the n students who participated in the experiment and nine columns corresponding to the nine responses requested in items 1-11 of the EXPERIMENTAL PROTOCOL.

In questions 9.1 - 9.7, you will examine the data to assess whether the randomization led to any imbalances regarding height, weight, GPA, sex, jock-status and smoking habits. In questions 9.9-9.10, you will examine the primary question of interest. Namely, whether there is any evidence in the data to suggest that the two approaches to the measurement of pulse yield different results. In question 9.11, you will follow-up on this analysis by comparing the difference between initial and final measurements across methods. In questions 9.12 - 9.14, you will attempt to improve the precision of this comparison by *post-hoc pairing*. The remaining questions deal with the question of whether sex, physical activity or smoking habits appear to affect the change in pulse rate following physical activity.

9.1 What proportion of the coin tosses resulted in a head? Construct a 95% confidence interval for p, the probability of heads, based on your data via

$$\hat{p} \pm 1.96\sqrt{\hat{p}(1-\hat{p})/n} \ .$$

Does your 95% confidence interval include the value 0.5? What does that imply?

9.2 (a) Draw a stem-and-leaf plot of the weights of those people whose flip resulted in a head. Describe the plot - remember shape, location and spread! Draw a stem-and-leaf plot of the weights of those people whose flip resulted in a tail. Describe the plot - remember shape, location and spread! Compare the two plots.

(b) Test whether the difference between the two groups can be attributed to chance. If you need to use a t distribution, be sure to check that the data follow the normal distribution by looking at a histogram. Construct a 95% confidence interval for the difference. Does your interval include the value 0.0? If yes, what does that imply? If no, what does that imply?

9.3 Repeat question 9.2 for the height data.

9.4 Repeat question 9.2 for the GPA data.

9.5 Construct a $2x2$ frequency table summarizing the number of males who obtained heads, the number of men who obtained tails, the number of women who obtained heads and the number of women who obtained tails. Compute the odds ratio and the correlation coefficient for this table. Does the sex of the tosser and the outcome of the toss appear to be correlated?

9.6 Construct a $2x2$ frequency table summarizing the number of jocks (those who work out three or more times per week) who obtained heads, the number of non-jocks who obtained heads, the number of jocks who obtained tails and the number of non-jocks who obtained tails. Compute the odds ratio and the correlation coefficient for this table. Does the jock-status of the tosser and the outcome of the toss appear to be correlated?

9.7 Construct a $2x2$ frequency table summarizing the number of heavy smokers (2+ packs per week) who obtained heads, the number of non-heavy-smokers who obtained heads, the number of heavy smokers who obtained tails and the number of non-heavy-smokers who obtained tails. Compute the odds ratio and the correlation coefficient for this table. Does the smoking status of the tosser and the outcome of the toss appear to be correlated?

9.8 Based on your responses to 9.1-9.7, is there any evidence of imbalance in the data?

9.9 Draw a stem-and-leaf plot for the initial pulses measured via the RADIAL method. Describe the plot - remember shape, location and spread! Draw the corresponding plot for the initial pulses measured via the CAROTID method. Describe the plot - remember shape, location and spread! Compare the two plots.

(b) Test whether the observed difference in average initial pulse between the two methods can be attributed to chance. If you need to use a *t* distribution, be sure to check that the data follow the normal distribution by looking at a histogram. Construct a 95% confidence interval for the difference. Does your interval include the value 0.0? If yes, what does that imply? If no, what does that imply?

9.10 Repeat question 9.9 for the final pulse data. Based on your computations in 9.9 and 9.10, can you conclude that the two approaches to the measurement of the pulse yield different results?

9.11 For each student, compute the final pulse rate less the initial pulse rate. Compute the average of the difference for those people who used the *CAROTID method*. Compute the average of the differences for those people who used the *RADIAL method*. Test whether the difference between these two averages can be attributed to chance. If you need to use a *t* distribution, be sure to check whether the data seem to follow the normal distribution. Also, be sure to check for outliers in your data. Is there any evidence to suggest whether the two methods give different results? Specify a range of plausible values for which we are 95% confident will include the difference. Does your interval include the value 0.0? If yes, what does that imply? If no, what does that imply?

9.12 Pair each individual who used the *RADIAL method* with an individual who used the *CAROTID method* on the basis of height. In other words, for a

RADIAL method person whose height is 67 inches find a *CAROTID method* person whose height is also 67 inches. (You may run out of *CAROTID method* people if the proportion of heads in your sample is greater than 0.5. In such a case, ignore the remaining *RADIAL method* people. Also, you may not be able to obtain an exact match for each pair. Try, however, to minimize the discrepancies in heights between the members of a pair.) After you have formed the pairs, for each pair subtract the final CAROTID pulse from the final RADIAL pulse. Based on these pairwise differences, test whether the difference between the two approaches can be attributed to chance. If you use a *t* distribution, be sure to check whether the data follow the normal distribution and to check for outliers. Construct a 95% confidence interval for the difference. Do your results differ from those obtained from the unpaired analysis in 9.10? If yes, in what way do they differ? In light of your computations in this question and in 9.10, would you redesign the experiment to determine whether the RADIAL and CAROTID methods yield different results? If yes, how would you redesign the experiment?

9.13 Pair the students on the basis of weight and repeat problem 9.12.

9.14 Pair the students on the basis of initial pulse rate and repeat problem 9.12.

9.15 For those students that used the *RADIAL method,* plot the final pulse rate versus the initial pulse rate. Describe what you see in this scatter plot. In a second plot, plot the final pulse rate versus the initial pulse rate and indicate which points belong to people who smoke two or more packs per week. In a third plot, plot the final pulse rate versus the initial pulse rate and indicate which points belong to people who work out three or more times per week. In a fourth plot, indicate which points belong to women. In a fifth plot, indicate those people with a GPA in excess of 3.0. In each of these plots, is there any indication that the indicated points seem to cluster in a different manner than the other points in the plot? Interpret!

9.16 Repeat problem 9.15 for those students that used the *CAROTID method.*

9.7 MINITAB APPENDIX

When it comes to approximating the reference distribution by a normal distribution or by a *t* distribution, MINITAB is unsurpassed. Given a list of numbers in column C1 and a list of numbers in column C2, the command

> TWOSample test and a K percent confidence interval for the data in columns C1 and C2

or more simply TWOS K C1 C2 , will perform the Welch two-sample test and return the *p-value* for the alternative that there is a "difference". Given this command, MINITAB will also construct the K% confidence interval for the difference, based on the Welch formula.

Given a list of numbers in column C1 and a list of numbers in column C2, the command

> POOLed test and a K percent confidence interval for the data in columns C1 and C2

or more simply POOL K C1 C2 , will perform the pooled two sample test and return the *p-value* for the alternative that there is a "difference". Given this command, MINITAB will also construct the K% confidence interval for the difference, based on the pooled two-sample formula.

In the MINITAB APPENDIX for Chapter 6, it was discussed how one may draw WITHOUT REPLACEMENT a sample of size *n* from a hat with *m* tickets using the SAMPle command. In particular, if the tickets in the hat are numbered 1 to 100, and these 100 numbers are stored in column C1, then SAMP 5 C1 C2 will select five numbers at random WITH REPLACEMENT between 1 and 100, inclusive. The corresponding five tickets are our sample. To draw a sample WITH REPLACEMENT, the command sequence draw at

```
RANDom 5 observations and put in column C1;
INTEgers between 1 and 100.
```

or more simply

```
RAND 5 C1;
INTE 1 100.
```

will select five numbers at random WITH REPLACEMENT between 1 and 100, inclusive. The corresponding five tickets are our sample.

To randomize 100 subjects to one of two treatments via the random sample size randomization design (with a *fair* coin) use the commands draw at

```
RANDom 100 observations and put in column C1;
BINOmial distribution with n equal to 1 and p equal to 0.5.
```

or more simply

```
RAND 100 C1;
BINO 1 0.5.
```

This command sequence will generate a series of 100 0's and 1's. One can now go through the list of subjects assigning a subject to a treatment depending on the corresponding value in column C2.

To randomize 100 subjects to one of two treatments via the fixed sample size randomization design (with 50 subjects to treatment #1 and 50 subjects to treatment #2), store fifty 1's and fifty 0's in column C1. The command sequence

> SAMPle 100 observations without replacement from column C1 and put in C2

or more simply

> SAMP 100 C1 C2

will randomly permute the fifty 0's and 1's. One can now go through the list of subjects assigning a subject to a treatment depending on the corresponding value in column C2.

10A

Synonyms

Key Words: Two-Sample Problem; Qualitative Data; $2x\,2$ Frequency Table.

> We have become Antipodean in our scientific expectations. You believe in the God who plays dice, and I in complete law and order in a world which objectively exists, and which I, in a wildly speculative way, am trying to capture.
>
> -Albert Einstein to Max Born (1944)

10A.1 INTRODUCTION

This chapter presents a brief discussion of the analysis of the $2x\,2$ frequency table. Having randomized a group of n subjects to one of two treatments, methodology to test for a treatment effect will be presented, as well as methodology to construct a range of plausible values for the difference. To illustrate this methodology, you will examine the effect of question wording on the outcome of a survey.

10A.2 THE $2x\,2$ FREQUENCY TABLE - RANDOMIZATION DESIGN

Assume that each of n subjects have been randomized to one of two treatments and a dichotomous response (trait present/trait absent) is recorded for each subject. The data are summarized as:

	Trait Present	*Trait* Not Present	Total
Treatment 1	a	b	a+b
Treatment 2	c	d	c+d
Total	a+c	b+d	

There are two approaches to the analysis of the data obtained in the randomization design. The *exact approach* (i.e. Fisher's Exact Test) is discussed in Brown and Hollander (1978), while the normal approximation is discussed in the following section.

10A.2.1 THE NORMAL APPROXIMATION

The normal approximation is quite simple. Given the above GENERIC $2x2$ table, let $\hat{p}_1 = \frac{a}{a+b} = \frac{a}{n_1}$ (the proportion of subjects *receiving treatment #1* who have the trait), let $\hat{p}_2 = \frac{c}{c+d} = \frac{c}{n_2}$ (the proportion of subjects *receiving treatment #2* who have the trait), and $\hat{p} = \frac{a+c}{a+b+c+d} = \frac{n_1\hat{p}_1 + n_2\hat{p}_2}{n_1+n_2}$ (the proportion of subjects *in the study* who have the trait). To test the null hypothesis that the two treatments are identical, one compares the statistic

$$\frac{\hat{p}_1 - \hat{p}_2}{\sqrt{\frac{\hat{p}(1-\hat{p})}{n_1} + \frac{\hat{p}(1-\hat{p})}{n_2}}} = \frac{\hat{p}_1 - \hat{p}_2}{\sqrt{\hat{p}(1-\hat{p})(\frac{1}{n_1} + \frac{1}{n_2})}}$$

to the normal distribution.

When is the normal approximation valid? One of the older rules-of-thumb (Cochran, 1954) for assessing the validity of the normal approximation, for a given GENERIC $2x2$ table, is to *not use* the normal approximation when one or more of the following quantities is less than 5.0:

$$\frac{(a+b)(a+c)}{a+b+c+d}, \frac{(a+b)(b+d)}{a+b+c+d}, \frac{(c+d)(a+c)}{a+b+c+d}, \frac{(c+d)(b+d)}{a+b+c+d}.$$

In addition to testing for a treatment effect, one may also be interested in constructing a confidence interval for the treatment effect. In other words, one may wish to specify a range of plausible values for the difference in proportions. In general, a $100\alpha\%$ confidence interval for the treatment effect, based on the normal approximation, is given by

$$\left(\hat{p}_1 - \hat{p}_2 - z\sqrt{\frac{\hat{p}_1(1-\hat{p}_1)}{n_1} + \frac{\hat{p}_2(1-\hat{p}_2)}{n_2}},\ \hat{p}_1 - \hat{p}_2 + z\sqrt{\frac{\hat{p}_1(1-\hat{p}_1)}{n_1} + \frac{\hat{p}_2(1-\hat{p}_2)}{n_2}}\ \right),$$

where z is the value such that the area under the normal distribution from $-z$ to $+z$ is equal to α.

It is important to point out that in testing one proceeds under the null hypothesis of no treatment effect. Thus, one uses the *pooled* estimate of the common proportion in the denominator of the test statistic. In constructing confidence intervals, one is not to use the estimate of the common proportion in computing the estimated standard error.

10A.3 MATERIALS

Each student should have a data sheet and a coin.

10A.4 EXPERIMENTAL PROTOCOL NAME____________________

DATA SHEET

1. Flip the coin. Record the result:

_________ HEAD _________TAIL

2. Record your sex:

_________ MALE _________FEMALE

3. Record your GPA: (A=4.0)

____

4. If you obtained a **HEAD,** then answer the following question:

VERSION #1

You are an Army officer caught in an ambush with 600 of your soldiers.

A: If you take route A, 200 of your soldiers will be saved.

B: If you take route B, 600 soldiers will be saved with a probability of one-third and none will be saved with a two-thirds probability.

Which route will you choose? ________A ________B

5. If you obtained a **TAIL,** then answer the following question:

VERSION #2

You are an Army officer caught in an ambush with 600 of your soldiers.

A: If you take route A, 400 of your soldiers will die.

B: If you take route B, no soldiers will die with a probability of one-third and 600 will die with a two-thirds probability.

Which route will you choose? ________A ________B

Hand in your results so the class results may be pooled.

10A.5 ANALYSIS

Your instructor will provide you with the class results, so that you may answer the following questions. You will be given five 2x2 tables. Table #1 will summarize the class results by treatment (version 1 or version 2) and by response (route A or route B). Tables #2 and #3 will break the data down by sex of the respondent. Table #4 will summarize the responses of individuals with a reported GPA greater than 3.0, while Table #5 summarizes the responses of those with a reported GPA of 3.0 or lower. In question 10A.1 you are to examine the two versions of the question posed in the previous section. In questions 10A.2-10A.6, you are to examine whether the question wording had an effect on the responses.

10A.1 Examine the two versions of the question. How do they differ? How are they similar?

10A.2 For the class results, test whether the proportion of "route A" responses varies with the version of the question. Construct a 95% confidence interval for the difference. Is 0.0 in your confidence interval? If yes, what does that imply? If no, what does that imply? Do the data pass Cochran's rule-of-thumb test? If not, try Fisher's exact test (see Brown and Hollander, 1978).

10A.3 Repeat question 10A.2 using the men's data.

10A.4 Repeat question 10A.2 using the women's data. Do the results seem to vary with the sex of the respondent?

10A.5 Repeat question 10A.2 using the GPA greater than 3.0 data.

10A.6 Repeat question 10A.2 using the GPA of 3.0 or below data. Do the results seem to vary with the GPA of the respondent?

10A.7 Based on your responses to 10A.2-6, what can you conclude about the relation between question wording and survey outcome? Does the relation appear to be sex or GPA related? (For a further discussion of this topic see McKean, 1985 or Kahneman, Slovic and Tversky, 1982.)

10A.6 REFERENCES

1. Brown, B. and Hollander, M. *Statistics: A Biomedical Introduction,* John Wiley, New York, 1977.

2. Cochran, W. G. "Some Methods for Strengthening the Common X^2 Tests." *Biometrics 10,* 417-451, 1954.

3. Kahneman, D., Slovic, P. and Tversky, A. *Judgement Under Uncertainty: Heuristics and Biases,* Cambridge University Press, Cambridge, 1983.

4. McKean, K. "Decisions", *Discover,* June, 1985.

10B

Congenial Generators

Key Words: Association; Independence; χ^2 Statistic; χ^2 Distributions; *r x c* Frequency Tables; Random Number Generators.

> If God has made the world a perfect mechanism, he has at least conceded so much to our imperfect intellect that, in order to predict little parts of it, we need not solve innumerable differential equations but can use dice with fair success.
>
> -Max Born (1948)

> Only God can make a random selection.
>
> -Marion Levy Jr.

10B.1 INTRODUCTION

This chapter presents the χ^2 statistic for use in the analysis of the *r x c* frequency table and other types of qualitative data. To illustrate some of the ideas presented in this chapter, each student will be asked to generate a stream of "random" numbers. Based on the resulting frequency tables, you will assess your classmates' abilities to work as random number generators.

10B.2 THE χ^2 DISTRIBUTIONS

Mathematically speaking, the χ^2 **distributions** are a family of distributions indexed by a positive integer parameter called the *degrees of freedom.* Thus, there is a χ^2 distribution for 1 degree of freedom, 2 degrees of freedom, ...

What follows is a table which specifies various areas under selected χ^2 distributions.

Areas to the Right Under Various χ^2 Distributions

df	0.95	0.90	0.70	0.50	0.10	0.05	0.025	0.01
1	0.004	0.02	0.15	0.45	2.71	3.84	5.02	6.64
2	0.10	0.21	0.71	1.39	4.60	5.99	7.38	9.21
3	0.35	0.58	1.42	2.37	6.25	7.82	9.35	11.34
4	0.71	1.06	2.20	3.36	7.78	9.49	11.14	13.28
5	1.14	1.61	3.00	4.35	9.24	11.07	12.83	15.09
6	1.64	2.20	3.83	5.35	10.65	12.59	14.45	16.81
7	2.17	2.83	4.67	6.35	12.02	14.07	16.01	18.48
8	2.73	3.49	5.53	7.34	13.36	15.51	17.53	20.09
9	3.33	4.17	6.39	8.34	14.68	16.92	19.02	21.67
10	3.94	4.86	7.27	9.34	15.99	18.31	20.48	23.21
11	4.58	5.58	8.15	10.34	17.28	19.68	21.92	24.73
12	5.23	6.30	9.03	11.34	18.55	21.03	23.34	26.22
13	5.89	7.04	9.93	12.34	19.81	22.36	24.74	27.69
14	6.57	7.79	10.82	13.34	21.06	23.69	26.12	29.14
15	7.26	8.55	11.72	14.34	22.31	25.00	27.49	30.58
16	7.96	9.31	12.62	15.34	23.54	26.30	28.85	32.00
17	8.67	10.09	13.53	16.34	24.77	27.59	30.19	33.41
18	9.39	10.87	14.44	17.34	25.99	28.87	31.53	34.81
19	10.12	11.65	15.35	18.34	27.20	30.14	32.85	36.19
20	10.85	12.44	16.27	19.34	28.41	31.41	34.17	37.57
21	11.59	13.24	17.18	20.34	29.62	32.67	35.48	38.93
22	12.34	14.04	18.10	21.34	30.81	33.92	36.78	40.29
23	13.09	14.85	19.02	22.34	32.01	35.17	38.08	41.64
24	13.85	15.66	19.94	23.34	33.20	36.42	39.36	42.98
25	14.61	16.47	20.87	24.34	34.38	37.65	40.65	44.31
30	18.49	20.60	25.51	29.34	40.26	43.77	46.98	50.89
40	26.51	29.05	34.87	39.34	51.81	55.76	59.34	63.69

How does this table work? Notice that the *rows* of the table correspond to *degrees of freedom* and the *columns* of the table correspond to *areas* under the various χ^2 distributions. A given value at row $df = d$ and under the column labeled $0.x$ yields the value such that *to the right of this value lies* $x\%$ *of the area of the* χ^2 *distribution on* d *degrees of freedom.*

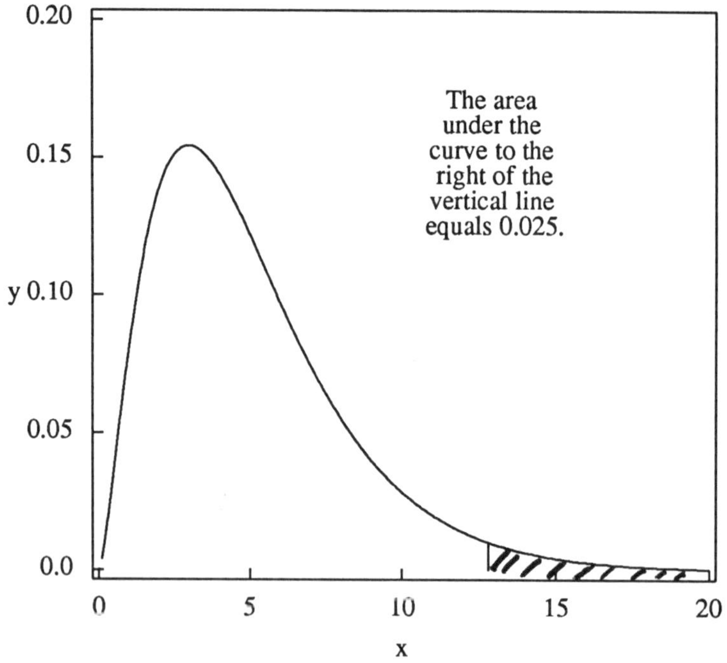

For example, the value 12.83 is found in the fifth row and the seventh column of the table. Note that the fifth row corresponds to a χ^2 distribution on 5 degrees of freedom, while the seventh column corresponds to an area under this distribution of 0.025. In this way, the table indicates that 2.5 percent of the χ^2 distribution on five degrees of freedom lies to the *right of 12.83*. Similarly, fifty percent of the χ^2 distribution on ten degrees of freedom lies to the right of 9.34.

10B.3 THE OMNIBUS χ^2 TEST

In this section, the use of the χ^2 statistic to test a variety of different hypotheses is illustrated. The commonality to all of these tests will be the comparison of the *observed* count to the *expected* count (the expected count being computed under the null hypothesis of interest) via the formula

$$\frac{(Observed\ Count - Expected\ Count)^2}{Expected\ Count}.$$

These tests will differ according to how one is to compute the *expected* count under the given hypothesis and according to the appropriate χ^2 distribution which one is to use as the reference distribution.

10B.3.1 TESTING A COMPLETELY DETERMINED NULL HYPOTHESIS

To motivate the χ^2 test for a completely determined null hypothesis, consider an experiment conducted by Gregor Mendel in 1863. In this experiment, Mendel had grown 529 plants and classified each of these plants into one of nine categories according to their seeds. Mendel's results are given in the following table:

Category	*Observed Count*
AABB	38
AaBB	60
aaBB	28
AABb	65
AaBb	138
aaBb	68
AAbb	35
Aabb	67
aabb	30

where A = smooth, a = wrinkled, B = yellow and b = green and aa represents a wrinkled pea; Aa, aA or AA represent a smooth pea; bb represents a green pea; bB, Bb or BB represent a yellow pea. According to Mendel's theory, the probability of obtaining each of these nine types of plants is given as:

Category	*Probability of Category Under Mendel's Laws*
AABB	1/16
AaBB	2/16
aaBB	1/16
AABb	2/16
AaBb	4/16
aaBb	2/16
AAbb	1/16
Aabb	2/16
aabb	1/16

The question of interest is whether the probabilities suggested by Mendel's laws and his data are in accord. In other words, we wish to test the completely determined null hypothesis that the probability of observing AABB is equal to $\frac{1}{16}$, the probability of observing AaBB is equal to $\frac{2}{16}$, ..., and the probability of observing aabb is equal to $\frac{1}{16}$.

How is one to implement this test? The idea is to examine the distance *between the data* and what one would *expect to have observed assuming Mendel's laws govern the inheritance of these characteristics* via the χ^2 statistic. The following four steps indicate how one is to measure this distance.

1. For each category, compute the *expected* number of plants assuming Mendel's laws to be correct.

2. Compute the *difference* between each observed count and the corresponding expected count.

3. *Square* these differences and divide each squared difference by the corresponding expected count.

4. *Sum* the terms computed in Step 3 over all categories.

To gauge the magnitude of the difference, compare the sum of step 4 to the appropriate χ^2 distribution.

Given that Mendel studied 529 plants, how many plants would he expect to fall in the AABB category? If his laws of inheritance are correct, he would expect to see $529(\frac{1}{16}) = 33.0625$ such plants. Similarly, he would expect to see $529(\frac{2}{16}) = 66.125$ plants fall in the AaBB category. The following table lists the observed and expected number of plants for each category.

Observed Count	*Expected Count*
38	33.0625
60	66.1250
28	33.0625
65	66.1250
138	132.250
68	66.1250
35	33.0625
67	66.1250
30	33.0625

If one then evaluates

$$\frac{(\textit{Observed Count} - \textit{Expected Count})^2}{\textit{Expected Count}}$$

for each of the nine categories and then sums these nine numbers, the χ^2 statistic is obtained:

$$0.73735 + 0.56734 + 0.77516 + 0.01914 + 0.25000 + 0.05316 + 0.11354 + 0.01157 + 0.28367 = 2.811 .$$

How is the statistical significance of this χ^2 statistic to be assessed? One compares this value to the χ^2 distribution on *eight* degrees of freedom.

> In the situation when the expected probabilities for each category is completely determined by the null hypothesis, for a *sufficiently large* sample size, one is to compare the χ^2 statistic to the χ^2 distribution with degrees of freedom specified by the following formula
>
> Number of Categories - 1 .

By *sufficiently large,* it is meant that all expected cell counts should be greater than 5.0.

For Mendel's data, the expected probability for each of the nine categories is determined by Mendel's laws. Thus, one can use the above rule to compute the correct degrees of freedom. In this regard, note that for Mendel's data there are nine categories. Thus, according to the rule, one is to compare the χ^2 statistic to the χ^2 distribution on eight degrees of freedom.

Locating the eighth row in the χ^2 table (corresponding to eight degrees of freedom), one finds that the area to the right of 2.811 under the χ^2 distribution on eight degrees of freedom is equal to a value which is somewhat less than 0.95. Thus, the *p-value* to test whether Mendel's law is congruent with the data is slightly less than 0.95. In this way, based on the above data, there is little evidence against the null hypothesis that the probability of each of the categories are distributed according to Mendel's laws. (See Freedman, Pisani and Purves, 1978 for a further discussion of these data.)

10B.3.2 THE *rxc* FREQUENCY TABLE - TESTING FOR EQUAL PROPORTIONS FOR r CATEGORIES

Suppose one draws a sample of size n_1 from hat #1 and a sample of size n_2 from hat #2, where in each hat there are four types of tickets. For example, assume we did a survey with possible responses "favorable", "not favorable", "not heard of" and "undecided". Thus, there are four types of tickets. To proceed, code *favorable* opinion as "1", *not favorable* opinion as "2", *not heard of* as "3" and *undecided* as "4".

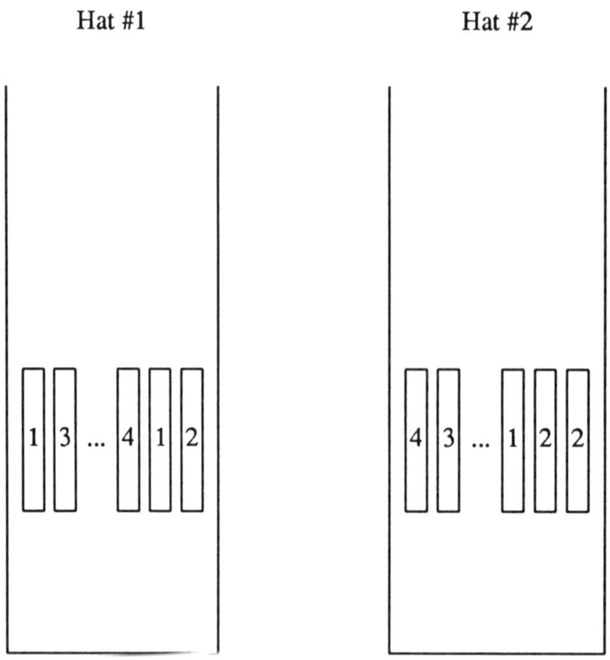

Having drawn a random sample of size n_1 from hat #1 and a random sample of size n_2 from hat #2, where each ticket in both hats is labeled with one of four possible labels, one can summarize the data in the following GENERIC $2x\,4$ frequency table:

	Sample #1	*Sample #2*
Label #1	a	b
Label #2	c	d
Label #3	e	f
Label #4	g	h
Total	n_1	n_2

Based on these data, it is of interest to test the null hypothesis that

the proportion of tickets with label #1 is identical for hat #1 and for hat #2

and

that the proportion of tickets with label #2 is identical for hat #1 and for hat #2

and

that the proportion of tickets with label #3 is identical for hat #1 and for hat #2 *and*

that the proportion of tickets with label #4 is identical for hat #1 and for hat #2.

(Note that this null hypothesis does not specify specific values for the various proportions. As such, it is *not* a completely determined null hypothesis.)

Under this null hypothesis, what is the *expected* count for each of the cells in the table? Under the assumption that the proportion of tickets with label #1 is identical for both hats, it seems sensible to *estimate* this common proportion as $\frac{a+b}{n_1+n_2}$. Thus, of the n_1 tickets in sample #1, one would expect $\frac{(n_1)(a+b)}{n_1+n_2}$ tickets to have label #1. In addition, of the n_2 tickets in sample #2, one would expect $\frac{(n_2)(a+b)}{n_1+n_2}$ tickets to have label #1. Similarly, under the assumption that the proportion of tickets with label #2 is identical for both hats, it seems sensible to *estimate* this common proportion as $\frac{c+d}{n_1+n_2}$. Thus, of the n_1 tickets in sample #1, one would expect $\frac{(n_1)(c+d)}{n_1+n_2}$ tickets to have label #2. In addition, of the n_2 tickets in sample #2, one would expect $\frac{(n_2)(c+d)}{n_1+n_2}$ tickets to have label #2. Continuing in this way, the following table of expected values is obtained:

	Sample #1	*Sample #2*
Label #1	$\frac{(n_1)(a+b)}{n_1+n_2}$	$\frac{(n_2)(a+b)}{n_1+n_2}$
Label #2	$\frac{(n_1)(c+d)}{n_1+n_2}$	$\frac{(n_2)(c+d)}{n_1+n_2}$
Label #3	$\frac{(n_1)(e+f)}{n_1+n_2}$	$\frac{(n_2)(e+f)}{n_1+n_2}$
Label #4	$\frac{(n_1)(g+h)}{n_1+n_2}$	$\frac{(n_2)(g+h)}{n_1+n_2}$
Total	n_1	n_2

In this way, for each observed count in the $2x4$ frequency table, one can evaluate the term

$$\frac{(Observed\ Count - Expected\ Count)^2}{Expected\ Count},$$

take the sum over all these quantities and then compare this sum to the appropriate χ^2 reference distribution, when the sample size is *sufficiently large*. (By *sufficiently large* it is meant that the *expected* counts for all cells in the table are all greater than 5.0.)

What is the appropriate χ^2 reference distribution? To test the null hypothesis of equal proportions for the *rxc* table, one compares the χ^2 statistic to the χ^2 distribution on $(r-1)(c-1)$ degrees of freedom.

In general, to compute the expected count for the ith row and the jth column in a *rxc* frequency table (to test the null hypothesis of equal proportions), one is to evaluate the formula

$$\frac{(i^{th}\ Row\ Total)(j^{th}\ Column\ Total)}{Total\ Number\ of\ Observations}.$$

One then computes

$$\frac{(Observed\ Count - Expected\ Count)^2}{Expected\ Count}$$

for each cell in the *rxc* table and then sums these values. The resulting χ^2 statistic is to be compared to the χ^2 distribution on $(r-1)(c-1)$ degrees of freedom.

10B.3.3 THE *rxc* FREQUENCY TABLE- TESTING FOR INDEPENDENCE

Consider a hat, where on each ticket in the hat is written two pieces of information. Also assume that the first piece of information (A) may be one of c possible types and the second piece of information (B) may be one of r possible types. Let A_1 denote the event

{The selected ticket is labeled with the 1st A type.} ,

A_2 denote the event

{The selected ticket is labeled with the 2nd A type.} ,

and so on until A_c which denotes the event

{The selected ticket is labeled with the cth A type.} .

Similarly, let B_1 denote the event

{The selected ticket is labeled with the 1st B type.} ,

B_2 denote the event

{The selected ticket is labeled with the 2nd B type.} ,

and so on until B_r which denotes the event

{The selected ticket is labeled with the rth B type.} .

To show that the two pieces of information on each ticket are independent, one must show that the event A_1 is independent of each of the events $B_1, B_2, ..., B_r$, the event A_2 is independent of each of the events $B_1, B_2, ..., B_r$, ..., and that the event A_c is independent of each of the events $B_1, B_2, ..., B_r$.

To construct a test, one is to proceed under the null hypothesis and then examine the distance between the data and how the data should look under the null hypothesis. Under the null hypothesis of independence, the expected count in the cell corresponding to the event

$$\{A_1 \text{ and } B_1\}$$

is given by

$$n\hat{p}(A_1)\,\hat{p}(B_1) = \frac{(\textit{First Row Total})(\textit{First Column Total})}{n},$$

where n is the total number of tickets *drawn from the hat*. Similarly, under the null hypothesis of independence, the expected count in the cell corresponding to the event

$$\{A_i \text{ and } B_j\}$$

is given by

$$n\,\hat{p}(A_i)\,\hat{p}(B_j) = \frac{(i^{th}\ \textit{Row Total})(j^{th}\ \textit{Column Total})}{n}.$$

Thus, the χ^2 statistic is obtained by evaluating the term

$$\frac{(Observed\ Count - Expected\ Count)^2}{Expected\ Count}$$

for all the cells in the *rxc* frequency table and then summing these *rc* terms, where *Observed Count* is the observed count in the ith row and j^{th} column of the *rxc* frequency table and *Expected Count* is the corresponding expected count given by the formula

$$\frac{(i^{th}\ Row\ Total)(j^{th}\ Row\ Total)}{n}.$$

What is the correct reference distribution for this χ^2 statistic? To test the null hypothesis of independence based on the data in a *rxc* frequency table, the correct reference distribution for the χ^2 statistic is the χ^2 distribution on $(r-1)(c-1)$ degrees of freedom, when the sample size is *sufficiently large*. By *sufficiently large* it is meant that the expected cell counts are larger than 5.0.

10B.4 RANDOM NUMBERS

What is a "stream of random numbers"? This question turns out to be one of those questions where it is easy to state an answer, but it is very hard to state an answer which can withstand close scrutiny.

As a potential answer to the question of "what is a stream of random numbers?", try this answer:

> A stream consisting of, for example, the digits 1, 2, and 3 is random if, given all but one digit in the stream, there is no formula which can predict the missing digit with a probability of better than $\frac{1}{3}$.

Assume that you flipped a coin four times and obtained four heads: 1, 1, 1, 1 (where "1" = head). It seems that such a stream of numbers should be called *a random stream,* since the stream was produced by a random mechanism, yet there is clearly a very simple formula which can be used to successfully predict any missing number in the series. Thus, according to the above definition, this stream of numbers would not be called a "random stream".

Is the following stream of numbers random?

7 4 8 0 6 3 1 9 5 2

That stream appears to be a random arrangement of the digits 0, 1, ..., 9. Gardner (1975), however, points out that there is a very simple rule which defines this stream.

> Start at the left and spell *zero*, counting one digit for each letter. The spelling ends on 0. Cross it out. Continue with *one*. Cross out 1. Proceed in this manner, spelling the digits that have not already been crossed out. The series is circular, if a count is uncompleted at the end of the line of digits, go back to the beginning. The arrangement makes it possible for all 10 digits to be spelled in numerical order.

According to the definition of "a stream of random numbers" currently under consideration, this curious stream was random until the trick was revealed. The stream then became "nonrandom". To some people, the characteristic of "randomness" should not be a function of time (until one determines the "rule") or a function of the observer. (I know the "rule", so to me the stream is not random - you do not know the rule, so for you the stream is random.)

Another approach, though also not without problems, is somewhat more objective.

> A stream of numbers is *pseudo-random, with respect to n tests,* if such a stream passes all n tests.

Clearly, one can imagine a stream of numbers which is pseudo-random with respect to n tests but not with respect to the $(n+1)^{st}$ test. Moreover, what one considers to be pseudo-random depends on how well the n tests characterize this person's sense of randomness. Clearly, such a sense of randomness may vary from person-to-person. Moreover, the question of which n tests are to be used to define pseudo-randomness is up for debate. We will adopt this test-based approach to define pseudo-randomness, since at the minimum one can rigorously define tests for the stream of pseudo-random numbers. The consumer of a stream of pseudo-random numbers can then examine which tests the stream of numbers pass, to decide if the stream is sufficiently random for his or her purposes. Of course, a problem with such an approach is that one does not know if there is some very simple statistical test, yet to be developed, for which the stream of numbers does not pass.

One approach to generate a stream of random numbers is to use a physical device, such as a coin or a spinner. In fact, the book entitled *A Million Random Digits with 100,000 Normal Deviates,* prepared by the Rand Corporation and published in 1955, reports a stream of one million random digits produced by an "electronic roulette wheel". In particular, a random frequency pulse source produced a stream of random binary digits which were converted to decimal digits. Of course, *a-priori* the developers of this stream did not know whether the stream would, in fact, be random. For this reason, they tested the stream using several statistical tests to insure that the stream of one million digits is, at least, pseudo-random.

Another approach is von Neumann's "middle of the square" method. His idea is to start with a number with *n* digits, square this number, and then take the middle *n* digits as the updated pseudo-random number. It is noted that this method can produce some clearly nonrandom pseudo-random numbers. For example, the square of 3792 is 14379264. Thus, starting with 3792, von Neumann's appoach leads to the stream 3792, 3792, 3792, ...

A somewhat less problematic arithmetic approach for generating pseudo-random numbers is that of the *linear congruential generator*. Mathematically speaking, the linear congruential generator is a specific type of function which maps integers into integers. In particular, given the integers a, c, m and the current pseudo-random number x_i, one obtains the next pseudo-random number x_{i+1} via the formula

$$x_{i+1} = ax_i + c \text{ (modulo } m) .$$

(The expression

$$u = s \text{ (modulo t)} ,$$

where s and t are integers, is shorthand for the following recipe:

Divide s by t and let u equal the integer remainder.

For example, 2 = 52 (modulo 10) because $\frac{52}{10}$ equal 5 plus a remainder of 2.) Given this modulo formula for generating pseudo-random numbers, one can "plug-in" a starting number or seed x_0 (into x_i) to obtain x_1. One can then "plug-in" x_1 (into x_i) to obtain x_2 and plug in x_2 to obtain x_3 and so on. For example, if $a = 999$, $c = 777$, $m = 666$ and $x_i = 23$, then x_{i+1} would equal

$$(999)(23) + 777 \text{ (modulo 666)} = 23754 \text{ (modulo 666)} = 444$$

and x_{i+2} would equal

$$(999)(444) + 777 \text{ (modulo 666)} = 444333 \text{ (modulo 666)} = 111.$$

So the stream of pseudo-random numbers would begin as 444, 111, ... The arithmetic of linear congruential generators is mindless, a perfect task for a computer.

There is a vast literature which discusses the mathematical properties of these linear congruential generators and what are the best values of a, c and m for a particular computer. In fact, the theory of linear congruential generators makes use of ideas in a branch of mathematics called number theory. (See Knuth, 1969 for an excellent review of this area.) The linear congruential generator has the advantages of being easy to implement on a computer, runs quite fast on a computer, does not require much space on a computer and the stream of numbers can be reconstructed given the seed x_0.

Of course, the output of a linear congruential generator is not at all random, since each element in the stream can be predicted exactly using the modulo formula. However, the output of the linear congruential generator has been shown to pass a number of statistical tests. (The linear congruential generator is, however, not infallible - see Marsaglia, 1968.)

What kind of tests do people use to test the output of pseudo-random number generators? One simple test is the *frequency test.* In this test, one examines whether the numbers in the stream are distributed as expected. For example, if one is generating pseudo-random numbers, each of which are equally likely to be 0, 1, 2, ..., or 9, then it is important to examine whether, in fact, the mechanism is generating digits which are equally likely to appear. For example, if one is generating streams of numbers, each number being one of the digits 0, 1, ..., or 9, then one expects about one-tenth of the digits to equal 0, about one-tenth of the digits to equal 1, ..., and about one-tenth of the digits to equal 9.

To illustrate how one would test this *equally likely* hypothesis, a linear congruential generator was programmed (with a = 1366, m = 714025 and c = 150889) to produce a stream of ten thousand pseudo-random numbers, each number being 0,1,2, ..., or 9. The following table summarizes the number of times each of the digits 0 through 9 occurred in the stream:

Digit	*Frequency of Occurrence*
0	991
1	1033
2	953
3	946
4	1003
5	992
6	1035
7	1039
8	1036
9	972

Is there any evidence to suggest that some of the digits are not equally likely?

Given that ten thousand numbers were generated, the expected frequency for each of the ten digits is one thousand. Of course, in practice, one would not expect to see exactly one thousand 0's, one thousand 1's ..., and one thousand 9's in this series, for that would be too unlikely given chance variation. The idea is to test whether the observed departures from one thousand are more extreme than *expected by chance.*

How is one to perform the test? Note that we have a completely determined null hypothesis, i.e. the Section 10B.3.1 results are applicable. Thus,

$$\frac{(Observed\ Count - Expected\ Count)^2}{Expected\ Count}$$

will be evaluated for each of the ten categories. In particular, we will subtract 1000 from each of the observed counts in the above table, square each of these diffferences, weigh each of the squared differences by $\frac{1}{1000}$ and then sum these quantities to obtain the χ^2 statistic. In this way, for the data in the above table, the χ^2 statistic is equal to

$$\frac{(991-1000)^2}{1000} + \frac{(1033-1000)^2}{1000} + \frac{(953-1000)^2}{1000} + \frac{(946-1000)^2}{1000} + \frac{(1003-1000)^2}{1000} +$$
$$\frac{(992-1000)^2}{1000} + \frac{(1035-1000)^2}{1000} + \frac{(1039-1000)^2}{1000} + \frac{(1036-1000)^2}{1000} + \frac{(972-1000)^2}{1000} = 11.19.$$

What is the appropriate reference distribution for this χ^2 statistic? Given that each expected count is much greater than 5.0, one can use one of the χ^2 distributions as the reference distribution. Note that the expected probability of a given digit at any point in the stream is completely determined by the null hypothesis of equally likely to be $\frac{1}{10}$. In this way, the degrees of freedom formula of Section 10B.3.1 indicates that the correct degrees of freedom is ten less one or nine. The area to the right of 11.19 under the χ^2 distribution on nine degrees of freedom, based on the χ^2 table is somewhere between 0.1 and 0.5. Thus, the *p-value* to test the equally likely hypothesis is somewhere between 0.1 and 0.5. In this way, there is very little evidence against the null hypothesis of equally likely digits.

Another test to which one should subject a stream of numbers is the *test of independence*. If one were to generate a stream of pseudo-random numbers, then one may want the numbers in the stream to be independent of one another.

How is one to test the hypothesis of independence? To begin, one can generate a series of pairs of pseudo-random numbers and ask whether the first element in the pair is independent of the second element of the pair. To illustrate, a linear congruential generator was programmed (with the same a, m and c as used above to illustrate the equally likely hypothesis) to produce ten thousand pairs of pseudo-random numbers, where each number in the pair can be 1, 2 or 3. The following table summarizes the number of times 1 was observed in the first element of the pair *and* 1 was observed in the second element of the pair; the number of times 1 was observed in the first element of the pair *and* 2 was observed in the second element of the pair; ...; the number of times 3 was observed in the first element of the pair *and* 3 was observed in the second element of the pair:

Counts of Ordered Pairs of Numbers

First # / Second #	*1*	*2*	*3*
1	1096	1133	1164
2	1098	1140	1037
3	1111	1097	1124

From this table, it is seen for example, that in 1097 instances the first element of the pair was "3" and the second element of the pair was "2". To these data, one can apply the χ^2 test of independence detailed in Section 10B.3.3. In particular, since the expected values for these data *under the hypothesis of independence* are given by

Expected Counts of Ordered Pairs of Numbers

First # / Second #	*1*	*2*	*3*
1	1121.4	1143.4	1128.2
2	1082.4	1103.7	1088.9
3	1101.2	1122.9	1107.9

one can evaluate the expression

$$\frac{(Observed\ Count - Expected\ Count)^2}{Expected\ Count}$$

for each cell in the table and then sum these values to obtain the χ^2 statistic

$$0.57 + 0.10 + 1.14 + 0.23 + 1.20 + 2.48 + 0.09 + 0.60 + 0.23 = 6.64\ .$$

From Section 10B.3.3 it is known that the reference distribution for this statistic is the χ^2 distribution on four [(3-1)(3-1)] degrees of freedom. From the χ^2 table, the area under the χ^2 distribution on four degrees of freedom to the right of 6.64 is somewhere between 0.1 and 0.5. Thus, the *p-value* for the test of independence is somewhere between 0.1 and 0.5. In this way, the data present very little evidence against the null hypothesis of independence.

Yet another test which one can use to examine a stream of numbers is the *hi and lo* test. In this test, one examines whether a high number is followed by a lower one as often as a lower number is followed by a higher one. For example, given the ten thousand pairs of numbers presented above

Counts of Ordered Pairs of Numbers			
First # / Second #	*1*	2	3
1	1096	1133	1164
2	1098	1140	1037
3	1111	1097	1124

one can ask whether a "1" is as equally likely to be followed by a "2" as by a "1" or a "3", we can ask whether a "2" is as equally likely to be followed by a "2" as by a "1" or a "3", and whether a "3" is as equally likely to be followed by a "2" as by a "1" or a "3". Under this hypothesis, it is assumed that all of the nine possible outcomes presented in the above table are equally likely. That is, each of these nine outcomes has a probability of $\frac{1}{9}$ of occurring. Given that one generated ten thousand such pairs, the expected value for each cell is $\frac{10000}{9}$ or 1111.111... . In this way, to peform the hi and lo test, one plugs in the nine observed counts and the expected counts (all equal to 1111.1111...) into the expression

$$\frac{(Observed\ Count - Expected\ Count)^2}{Expected\ Count}$$

and then sums these values to obtain

$$0.21 + 0.43 + 2.52 + 0.15 + 0.75 + 4.94 + 0.00 + 0.18 + 0.15 = 9.33.$$

The χ^2 distribution is used to assess the statistical significance of the χ^2 statistic of 9.33. Because the null hypothesis states that all nine outcomes are equally likely, the expected probability of a given cell (to be $\frac{1}{9}$) is completely determined. Thus, the degrees of freedom formula of Section 10B.3.1 is used to indicate that the correct degrees of freedom is nine less one or eight. The area to the right of 9.33 under the χ^2 distribution on eight degrees of freedom, based on the χ^2 table, is somewhere between 0.1 and 0.5. Thus, the *p-value* for the hi and lo test is somewhere between 0.1 and 0.5. In this way, there is very little evidence in the data to suggest that the linear congruential generator does not pass the hi-lo test.

10B.5 MATERIALS

Each student should have a data sheet.

10B.6 EXPERIMENTAL PROTOCOL NAME________________________

DATA SHEET

1. Close your eyes and think of an integer between 1 and 3, inclusive. Record the value:

   ________ **1** ________ **2** ________ **3**

2. Close your eyes and think of an integer between 1 and 3, inclusive. Record the value:

   ________ **1** ________ **2** ________ **3**

3. Close your eyes and think of an integer between 1 and 3, inclusive. Record the value:

   ________ **1** ________ **2** ________ **3**

4. Record your sex.

   ________ **Male** ________ **Female**

Hand in your results so the class results may be pooled.

10B.7 ANALYSIS

Your instructor will provide you with the class results, so that you may answer the following questions. You will be given a rectangular array of numbers, with n rows corresponding to the n students who participated in the experiment and *four* columns corresponding to the three rounds of digit generating and the sex of the generator ("1"=male and "2"=female).

10.1 For the first round of digit generating, assess whether the digits 1, 2, and 3 were generated in an *equally likely* manner. Is there evidence to suggest that the digits were not generated in an *equally likely* manner?

10.2 Repeat question 10.1 using the data of the second round.

10.3 Repeat question 10.1 using the data of the third round.

10.4 Pool all three rounds and test whether the digits 1, 2, and 3 were generated in an *equally likely* manner. Based on your responses to questions 10.1-10.4, did the students in your class have difficulty in generating the digits in an *equally likely* manner?

10.5 Compare the responses of the men and the women based on the data in the first round. (Hint: construct a $2x3$ frequency table.) Is there any evidence to suggest that men and women differ with regard to random digit generation? If yes, discuss how they differ - does one group show a preference to a given digit?

10.6 Repeat question 10.5 using the data of the second round.

10.7 Repeat question 10.5 using the data of the third round.

10.8 Construct a $3x3$ frequency table based on the results of the first and second rounds of digit generating. (For example, the upper left cell in the $3x3$ table represents the number of instances in which a "1" was selected on the first and the second round.) Test whether each of the nine possible outcomes is *equally likely*. Test whether the results of the first and second round are *independent*.

10.9 Repeat question 10.8 using the data of the second and third rounds.

10.10 Repeat question 10.8 using the data of the first and third rounds. Based on your responses to questions 10.1-10.10, is there any evidence that the students in your class had difficulty in generating the digits in an *independent* manner?

10.11 Repeat questions 10.8-10.10 using the numbers generated by the men.

10.12 Repeat questions 10.8-10.10 using the numbers generated by the women. Discuss any differences you observe between the numbers generated by the men and by the women.

10B.8 MINITAB APPENDIX

Given an *rxc* table, MINITAB excells in computing the χ^2 statistic for the tests of *independence* and of equal proportions. In particular, if the first column of the *rxc* table is stored in column C1, the second column of the *rxc* table is stored in column C2, ..., the cth column of the *rxc* table is stored in column Cc, then the command compute the

CHISquare of the data stored in columns C1-Cc

or more simply CHIS C1-Cc, will compute the expect counts

$$\frac{(i^{th}\ Row\ Total)(j^{th}\ Column\ Total)}{Total\ Number\ of\ Observations}$$

for all the cells in the table and sum the quantities

$$\frac{(Observed\ Count - Expected\ Count)^2}{Expected\ Count}$$

over all the cells in the table to obtain the χ^2 statistic.

MINITAB can also be used to obtain the *p-value* for the χ^2 test. If the observed value of the χ^2 statistic is X.Y, then the command compute the

CDF up to X.Y;
CHISquare on *d* degrees of freedom.

or more simply

CDF X.Y;
CHIS *d* .

will compute the area under the χ^2 distribution on *d* degrees of freedom **to the left of** X.Y. To obtain the *p-value* for the χ^2 test, simply subtract this value from 1.0.

MINITAB can also generate *pseudo-random* digits. For example, if we want a stream of one hundred numbers, each number *equally likely* to be 3, 4, ..., or 32, give MINITAB the command

```
RAND 100 numbers and put these numbers in column C1;
INTEgers selected from the numbers between 3 and 32 inclusive.
```

or more simply

```
RAND 100 C1;
INTE 3 32.
```

Finally, you will be given a $4xn$ rectangular array of numbers. For these data, you will be asked to summarize the first two columns of this array, corresponding to the results of the first and second rounds of random digit generating, using a $3x3$ frequency table. This can be done quite easily in MINITAB . If the data that you wish to summarize are stored in columns C1 and C2, then the command construct a

```
TABLe based on the data in columns C1 and C2
```

or more simply TABL C1 C2, will construct the summarizing frequency table.

10B.9 REFERENCES

1. Freedman, D., Pisani, R. and Purves, R. *Statistics* Norton, New York, 1978.

2. Gardner, M. *Mathematical Carnival* Alfred A. Knopf, New York, 1975.

3. Knuth, D. E. *The Art of Computer Programming. Vol. 2,* Addison-Wesley, Reading, 1969.

4. Marsaglia, G. "Random Numbers Fall Mainly in the Planes." *Proceeding of the National Academy of Sciences, 61,* 25-28, 1968.

11

The Great Debate

Key Words: Pie Chart; Bar Graph; Straight Line Regression; Regression Curve; Running Median; Least Squares; Slope; Intercept.

He stretched out the line.
-The Book of Lamentations

11.1 INTRODUCTION

The pie chart (Playfair, 1801) is a popular device for illustrating the breakdown of expenditures or debts. For example, in Chapter 2 we examined the relative allocation of money by the National Endowment for the Arts in 1983 to its various programs using the bar graph. Here is the corresponding pie chart:

Federal Aid to the Arts - 1983

Total Budget: 145.3 Million Dollars

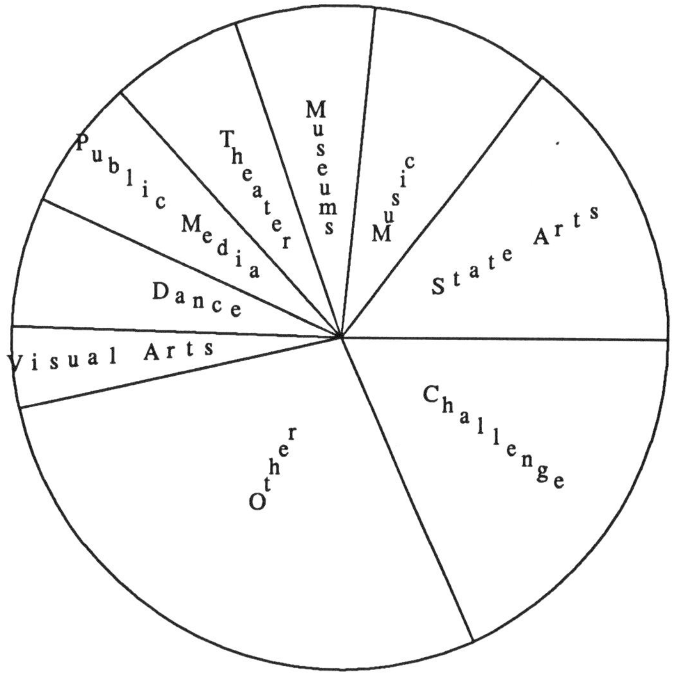

For the pie chart, the angle of a slice is proportional to the fraction which the slice represents. As can be seen, the pie chart, like the bar graph of Chapter 2, depicts the relative allocation of the National Endowment for the Arts budget to each program.

Some sixty years ago, a battle was fought regarding the relative merits of pie charts and bar graphs. On the one hand, were the pieniks who felt that pie charts could be read more *rapidly,* more *easily,* and more *accurately* than bar graphs. To quote the noted pienik, Walter Crosby Eells (1926):

> The use of circle diagrams to show component parts is worthy of encouragement; they should be recommended, not only on account of their popularity and psychological appeal, but also on the basis of scientific accuracy; their use should be considered a *"compliment to a man's intelligence."*

In contrast to this, K. G. Karsten (1923), a vocal barist, has commented regarding the pie chart:

> The disadvantages of the pie chart are many. It is worthless for study and research purposes. ... Before using it in place of the simpler and sounder 100 per cent bar, you should carefully gauge your audience or readers, and only if you believe you have begun to strain their interest should you judiciously insert the pie chart. In a sense, it might be construed as an *insult to a man's intelligence* to show him a pie chart.

Clearly, there is a strong difference of opinion between these two men regarding the merits of the pie chart and the bar graph. But how can one decide, one way or the other, which is preferable? First, one can outline a set of criteria with which to compare the two graphical techniques. The speed with which one can read the graph is an important criteria. However, speed is not enough, as the accuracy with which the graphs are read is also important. Yet another important criteria is mass appeal. If one uses a graphical technique which people find unappealing, then the reader may skip the graph and the point will be lost.

Having developed a set of criteria for comparing the pie chart and the bar graph, what would one do next? Well, one could argue deductively for one's favorite graphical technique. In a deductive argument, one would rely on personal opinion or experience to develop and support certain points. For example, Karsten (1923) argued that pie charts are bad because "the human eye cannot easily compare as to length the various arcs about the circle, lying as they do in different directions". However, the difficulty with this deductive approach is that it does not allow for the examination of the premises of the argument nor does it make an adequate investigation of the conclusions. In fact, Eells' (1926) rejoinder to Karsten was to point out that it is important to conduct "an experiment designed to test, by objective means, the validity of the criticisms". Thus, in conjuction with the deductive approach, one can perform an experiment, that is, take up the inductive approach to provide a basis for comparison. In regard to the great pie chart/bar graph debate, a number of people (Eells, 1926; Croxton, 1927; Croxton and Stryker, 1927; and Croxton and Stein, 1932) have performed experiments which compare the bar graph and pie chart with respect to such criteria as speed, accuracy and appeal, to assess which of the two graphical approaches is best.

More recently, Cleveland and McGill (1984a) have begun to develop a theory for the process of *graphical perception*. In their work, they identify a set of *elementary perceptual tasks* which are performed when a human attempts to extract information from a graph. Cleveland and McGill then order these tasks on the basis of the accuracy with which people perform these tasks. (The authors are quick to point out that accuracy is not the end all of graphics. Rather, the point of the authors is that one wants to avoid graphs which lead to disorganization and misconception.) Aside from having psychological importance, that is, presenting some insight into how humans decode graphical information, such a list provides a touchstone for the development of graphical methods. In particular, as Cleveland and McGill point out: "graphs should employ elementary tasks as high in the ordering as possible." If the extraction of information from a graph requires certain perceptual tasks which most people cannot do very well, then that graph is a bad graph. On the other hand, those graphs which do not necessitate atypical abilities to extract information are better.

One of the questions that Cleveland and McGill address is how to rank the elementary visual tasks of perception of *angle* and the perception of *position on a common scale*. By angle, it is meant how accurately does one perceive an angle between two intersecting lines? By position on a common scale, it is meant how accurately does one perceive the relative position of two points on the same line with respect to a common scale? Now notice that these two tasks are required in extracting information from the pie chart and from the bar graph. In the pie chart, an important elementary visual task for obtaining the numerical information is the perception of an angle. In the bar graph, judging the position of the top of the bars in relation to the vertical axis of the graph is required to extract the numerical information.

The focus of this chapter's investigation is to examine how people decode the information encoded in the graphs. Therefore, you will do a pie chart/bar graph experiment.

The notion of straight line regression will be presented from the perspective of least-squares. That is, given that a scatter plot seems to follow a straight line, what is the "best" line that one can draw through this point cloud? Next, the question of how can one smooth a scatter plot to obtain a regression curve will be discussed. These tools will then be used to analyze the data from the pie chart / bar graph experiment.

11.2 REGRESSION AS LEAST SQUARES

Given a scatter plot which seems to follow a straight line, it is of interest to determine the line which best fits this point cloud. Assume that the scatter plot is based on n pairs of numbers (x_1, y_1), (x_2, y_2), ..., (x_n, y_n). For these points, one wishes to find the values of γ and β in the equation of a straight line

$$y = \beta x + \gamma$$

which best fit the data.

One approach to the selection of γ and β is simply to "eye-ball" the plot and draw the line which seems best. One could then determine the best value of γ and β off of this "eye-ball" line. In some cases, the scatter about the line may be so small, that this approach may work quite well. However, given that the data do scatter about the line, it would be of interest to have an automated algorithm for selecting the "best" value of γ and β. It would also be important to *quantify* the precision of these "best" values.

To begin, define the residual e_i to equal y_i - $\beta\, x_i$ - γ. The residual, e_i, represents the distance between the *observed* response y_i and the value *predicted* by the model $\beta\, x_i + \gamma$. Consider the sum of the squared residuals

$$e_1^2 + e_2^2 + \ldots\ e_n^2 = (y_1 - \gamma - \beta x_1)^2 + (y_2 - \gamma - \beta x_2)^2 + \ldots\ (y_n - \gamma - \beta x_n)^2 \,.$$

The *principle of least-squares* indicates that one select the value of γ and β which makes this sum of squared residuals as small as possible. This sum represents the squared *distance* between the data and what is expected under the model. The smaller this distance, the closer the model is to the data. If one denotes the value of γ and of β which minimize this distance as $\hat{\gamma}$ and $\hat{\beta}$, respectively, then $\hat{\gamma}$ and $\hat{\beta}$ determine the straight line which is closest to the data.

Given the n pairs of numbers (x_1, y_1), (x_2, y_2), ..., (x_n, y_n), it turns out via a standard argument from second-year calculus, that $\hat{\beta}$ is equal to

$$\frac{(x_1-\bar{x})(y_1-\bar{y}) + (x_2-\bar{x})(y_2-\bar{y}) + \cdots + (x_n-\bar{x})(y_n-\bar{y})}{(x_1-\bar{x})^2 + (x_2-\bar{x})^2 + \cdots + (x_n-\bar{x})^2} = r\ \frac{SD_y}{SD_x}\ ,$$

where r is the correlation coefficient, SD_y is the standard deviation of the $y's$ and SD_x is the standard deviation of the $x's$. The quantity $\hat{\gamma}$ is equal to

$$\bar{y} - \hat{\beta}\,\bar{x}.$$

Given that the least-square estimates of γ and β, $\hat{\gamma}$ and $\hat{\beta}$, are determined given a sample of data, associated with each of these values is a certain degree of statistical imprecision. It is important to quantify the variability associated with these estimates.

When one uses least-squares to obtain $\hat{\beta}$, the estimated standard error of $\hat{\beta}$ is equal to

$$\sqrt{\frac{(y_1-\hat{\gamma}-\hat{\beta}x_1)^2+(y_2-\hat{\gamma}-\hat{\beta}x_2)^2+\cdots(y_n-\hat{\gamma}-\hat{\beta}x_n)^2}{n-2}}\ \sqrt{\frac{1}{[(x_1-\bar{x})^2+(x_2-\bar{x})^2+\cdots+(x_n-\bar{x})^2]}}\ ,$$

while the estimated standard error of $\hat{\gamma}$ is given by

$$\sqrt{\frac{(y_1-\hat{\gamma}-\hat{\beta}x_1)^2+(y_2-\hat{\gamma}-\hat{\beta}x_2)^2+\cdots(y_n-\hat{\gamma}-\hat{\beta}x_n)^2}{n-2}}\ \sqrt{\frac{1}{n}+\frac{\bar{x}^2}{[(x_1-\bar{x})^2+(x_2-\bar{x})^2+\cdots+(x_n-\bar{x})^2]}}\ .$$

One can construct a 100α% confidence interval for γ and for β if the following four assumptions are valid for the particular data set:

1. The scatter plot of the data follows a *straight* line.
2. The distribution of the errors follows the normal distribution.
3. The standard deviation of the errors is constant.
4. The errors are independent.

If these assumptions hold, then one can use

$$\hat{\gamma} \pm t\,(ESE\ of\ \hat{\gamma})$$

$$\hat{\beta} \pm t\,(ESE\ of\ \hat{\beta})\ ,$$

as 100α% confidence intervals for γ and β, respectively, where t is the value such that the area under the t distribution, on $n-2$ degrees of freedom, from $-t$ to $+t$ is equal to α. Johnson and Bhattacharyya (1985) present several graphical tools for determining whether these assumptions are valid for a particular data set.

11.3 THE REGRESSION CURVE

To illustrate the role of smoothing, consider the following data which represent the body temperature of a cow measured at 6:30 a.m. on thirty-four consecutive days. The temperature was measured using a telemetric thermometer which is implanted in the cow and sends radio "chirps" to a receiver. The following data are the number of chirps per minute on a given morning, where the higher the count, the higher the temperature. These data are taken from a report by Alba and Zartman (1979) and are discussed in detail in the book by Velleman and Hoaglin (1981).

60, 70, 54, 56, 70, 66, 53, 95, 70, 69, 56, 70, 70, 60, 60, 60, 50, 50,
48, 59, 50, 60, 70, 54, 46, 57, 57, 51, 51, 59, 41, 46, 40, 40 .

Plotting the count versus the day on which the observation was made, one obtains the following plot:

A Cow's Temperature Measured Over Thirty-Four Consecutive Days

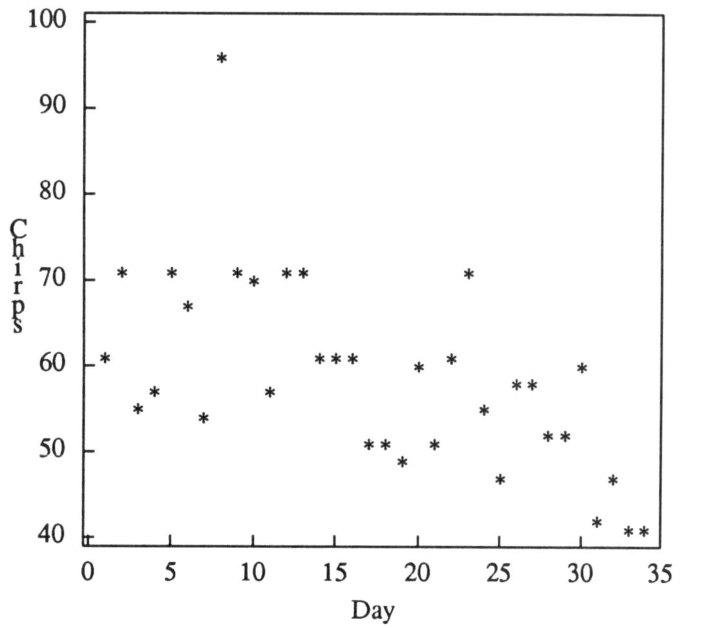

Note that there is an obvious outlying value on day 8. Given the extensive variability in the data, it may be difficult to detect any other sort of pattern in the scatter plot.

If, however, one looks hard at the plot, then it may appear that the number of chirps begins to decrease around day 14. What can one do to highlight any structure in these data?

To address this question, consider how one may smooth a scatter plot. One could partition the scatter plot into vertical strips, compute the "average point" within each strip and then pass a straight line through these "average points". That would be a reasonable approach if the data followed a straight line. However, for our chirp data we noticed a possible nonlinear pattern to the data - the pattern of the data appear to change in the third week. Fitting a straight line to the data would hide this structure. One of the basic themes of this book is the *appropriate* reduction of data. In reducing data, one runs the risk of hiding some important structure of the data. In the present situation of reducing a scatter plot to a recognizable pattern, one has to take great care not to oversmooth. That is, one wants to filter out the noise in the scatter plot, but not to distort the signal in the process.

There are a number of very flexible smoothers to highlight nonlinear structure in a plot. One class of smoothers is called **running medians.** At this point, it is instructive to first look at the *use* of a running median smoother. Following this presentation, we will discuss how the running median works. In particular, passing our chirp data through a running median smoother and then superimposing the resulting curve on the scatter plot one obtains:

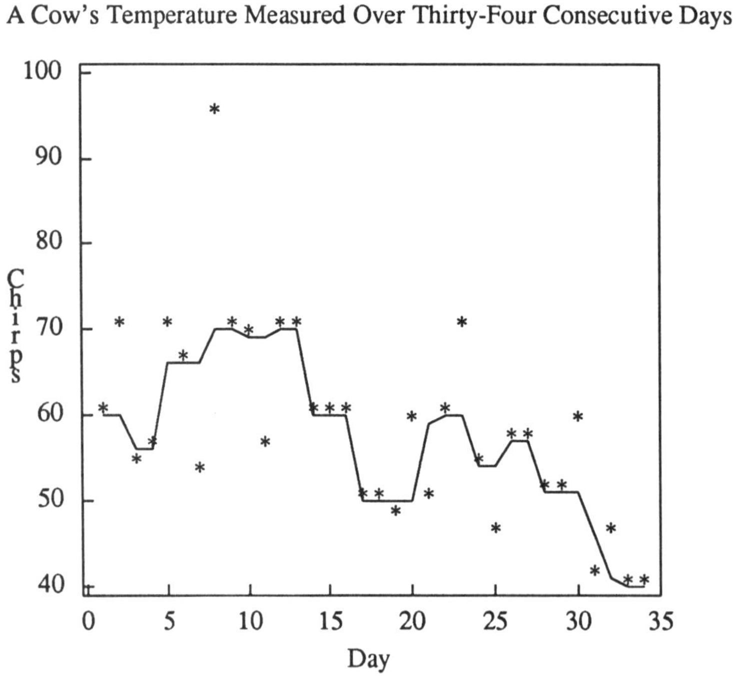

By superimposing the smoothed curve on the scatter plot, that is, filtering out the noise, one can detect a pattern in the data. In particular, during the thirty-four days, the cow's early morning temperature is seen to *cycle* with a periodicity of about 17 days. Over days 2-10 the smooth curve rises, while during the next 7 days it falls. The curve again rises from day 17 to day 23, and then begins to fall. Of course, this cyclical pattern is not unexpected. Cows, like humans, go through a reproductive cycle and the early morning temperature, in general, reflects this cycle. While there is significant variability to these data, one can see that by smoothing the chirps, there is an *overall pattern.*

Continuing the examination of the smoothed curve, one finds another feature of the data that requires comment - there is a steady decline in the chirp rate over the thirty-four days. One could speculate that the decline is due to the cow - possibly the change in season would affect the early morning temperature of the cow. Alternatively, the decline could be due to the fact that the batteries in the telemetric thermometer ran down over time.

For each of the points (x_i,y_i) on the scatter plot, we have plotted a new point $(x_i,\hat{y}_i)$. The x component of this new point is the same as the x component of the corresponding observed value, while the y component of this new point is a smoothed version of the old y_i. The curve which is traced out by the new points is called the *regression curve of y on x.*

How does one obtain $\hat{y}_i$, the smoothed value? Generally speaking, for each of the observed points (x_i, y_i), one looks at the data which "surround" this point and then "averages" this information to obtain the smoothed value $\hat{y}_i$. More particularly, for each point (x_i, y_i), one computes the median of those y values whose corresponding x value is in an interval about x_i. This median is then the smoothed value, which is called $\hat{y}_i$. By computing the median of the neighbors of y_i, one smooths out any y value which is very different from its neighbors. In this way, one property of a smooth curve is that there are no sudden changes in the shape of the curve.

To illustrate this smoothing process, let's compute $\hat{y}_i$ for day 8. To compute this smoothed value, one computes the median of y_7, y_8 and y_9, i.e. the median of 53, 95 and 70. The smoothed value, $\hat{y}_8$ is equal to 70.0. Thus, the seemingly discrepant y_i value of 95 is replaced by the number 70.

Each of the points in the plot is surrounded by a "window" and one then computes the median of the y values within this "window". We then move this "window", one point at a time, over all the points in the scatter plot to obtain the $\hat{y}_i$'s. These $\hat{y}_i$'s are then connected by straight lines to obtain the resulting smooth curve. By moving this window across the scatter plot and computing the median within the window one computes a series of *running medians.* For example, here is the "window" for smoothing y_8 :

A Cow's Temperature Measured Over Thirty-Four Consecutive Days

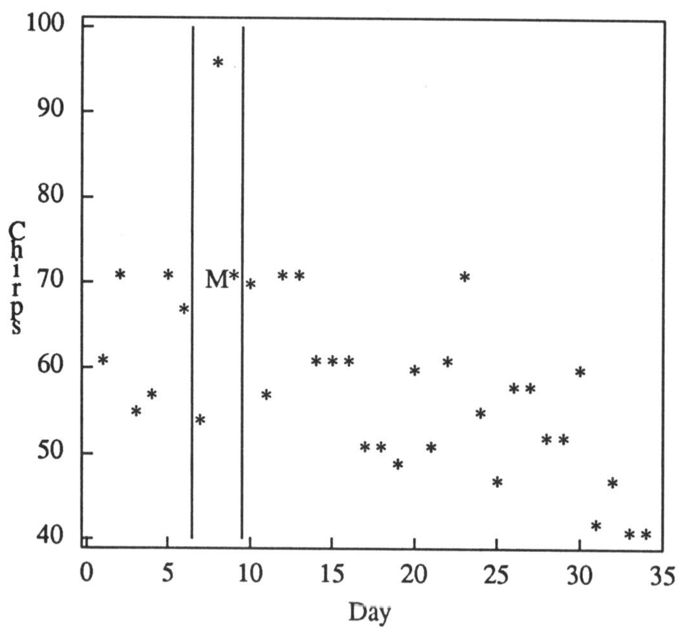

To obtain the smoothed value for day 8, one first looks at the number of chirps on day 7, day 8 and day 9 and then takes $\hat{y}_8$ to be the median of these 3 numbers. This "window" is moved from the left of the scatter plot to the right of the scatter plot, point by point, and the median for each point is computed. In this way, the resulting regression curve is traced out. (Note that the day 1 observation has no value to its left and the day 34 observation has no value to its right, that is, one cannot surround either of these two points by a "window". Hence, by convention, we let $\hat{y}_1 = y_1$ and $\hat{y}_{34} = y_{34}$.)

How wide should the moving window be? The width of the window depends on the nature of the data. The goal of the running median smoother is to filter out the noise, not to distort the signal. By making the window too small, one will not do a good job in filtering the noise. By making the window too large, one runs the risk of hiding some feature of the data. For the above data, we examined three observations in each window. Alternatively, we could have looked at five, seven... . It is a matter of trial and error to select the degree of smoothing which filters out the noise, but does not appear to distort the signal.

There are a wide range of smoothers that one can use to filter the scatter plot, of which the running median is probably the simplest. The paper by Cleveland and McGill (1984b) presents a survey of this methodology, as well as other techniques for highlighting the patterns in a scatter plot.

11.4 MATERIALS

Each student should have the six bar graphs and six pie charts of Section 11.5.

11.5 EXPERIMENTAL PROTOCOL

For each of the following six bar graphs and pie charts on pages 200-205, you will note that one bar in the bar graph and one wedge of the pie chart is marked with an X . What you are to do is to *estimate* what percent each of the other bars or wedges is of the largest bar or wedge. To illustrate, here is a pie chart and a bar graph.

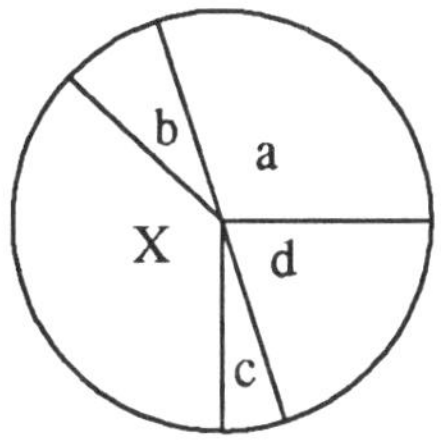

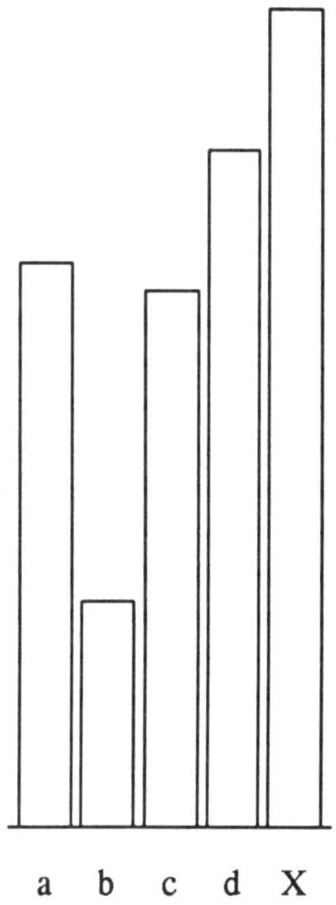

For the pie chart, I would estimate that the wedge labeled 'a' takes up about 80% of the area of the wedge labeled 'X', the wedge labeled 'b' takes up about 20% of the area of the 'X' wedge, the wedge labeled 'c' takes up about 15% of the area of the 'X' wedge and the wedge labeled 'd' takes up about 55% of the area of the 'X' wedge. Because I have access to the true percentages, I can compute the measurement errors *(estimated percent - true percent)* associated with these estimates: -1% for wedge 'a', -2% for wedge 'b', 2% for wedge 'c', and 1% for wedge 'd'. For the bar graph, I would estimate that the bar labeled 'a' takes up about 70% of the area of the 'X' bar, the bar labeled 'b' takes up about 25% of the area of the marked bar, the bar labeled 'c' takes up about 65% of the area of the marked bar, and the bar labeled 'd' takes up about 80% of the marked bar. The associated measurement errors are: 1% for bar 'a', -3% for bar 'b', -1% for bar 'c', and -2% for bar 'd'. That's all there is to it.

On the following six pages, are six pie charts and bar graphs. For each bar graph and pie chart, you are to *estimate* what percent each of the bars or wedges (marked a-d) is of the bar or wedge maked with an 'X'. Note that the pie chart and bar graph DO NOT represent the same percentages, that is, the bar graph and pie chart are not related. Also, for the purpose of this investigation, DO NOT stop to make any sort of measurements. All you have to do is to estimate the percentages with quick visual judgements.

FIRST PASS

For the pie chart, you are to judge what percent each wedge is of the wedge marked with an "X". For the bar graph, you are to judge what percent each bar is of the bar marked with an "X". The bar graph and pie chart are **NOT** related. **DO NOT** make any sort of measurements to obtain your estimates, just use a quick visual judgement. Record your results on the bottom of this page.

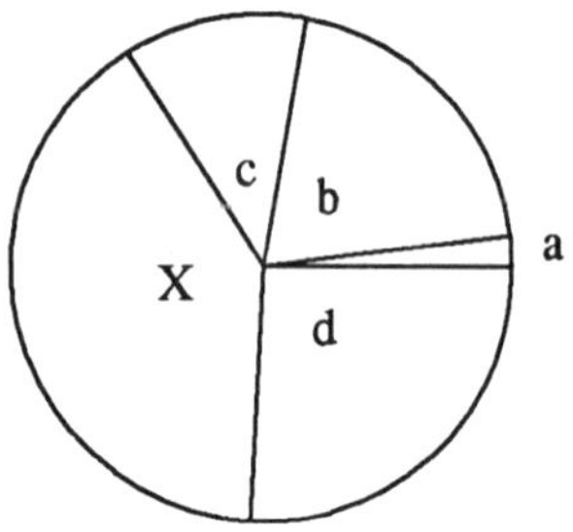

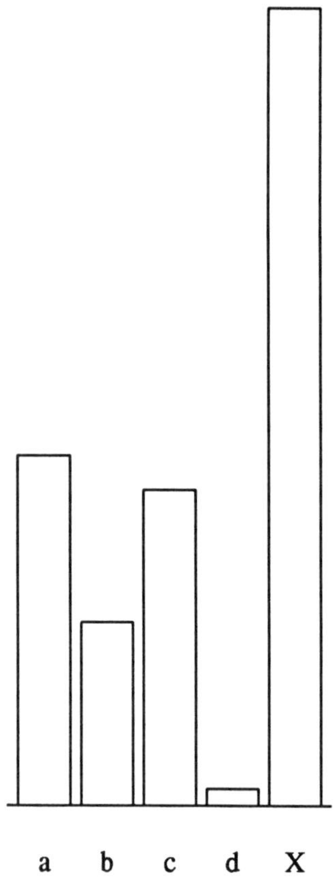

For the pie chart, my estimates (in percent) are:

(a/X) __________ (b/X) __________ (c/X) __________ (d/X) __________.

For the bar graph, my estimates (in percent) are:

(a/X) __________ (b/X) __________ (c/X) __________ (d/X) __________.

SECOND PASS

For the pie chart, you are to judge what percent each wedge is of the wedge marked with an "X". For the bar graph, you are to judge what percent each bar is of the bar marked with an "X". The bar graph and pie chart are **NOT** related. **DO NOT** make any sort of measurements to obtain your estimates, just use a quick visual judgement. Record your results on the bottom of this page.

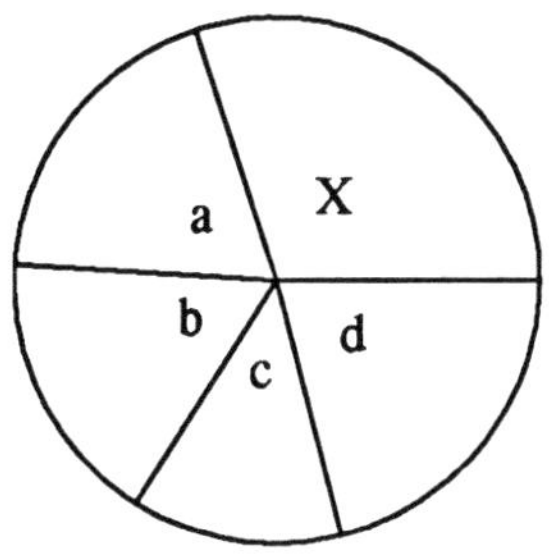

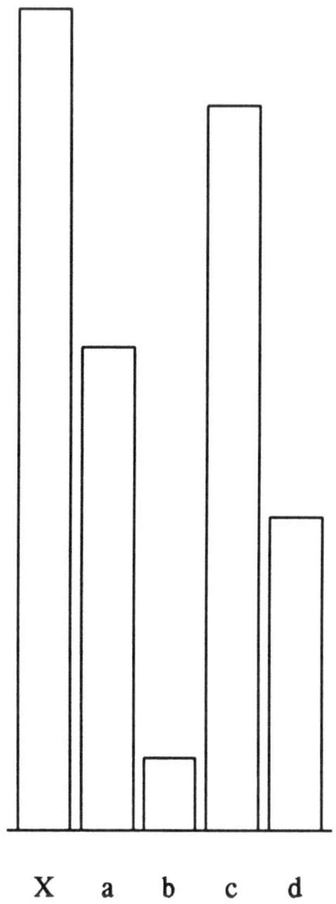

For the pie chart, my estimates (in percent) are:

(a/X) _________ (b/X) _________ (c/X) _________ (d/X) _________.

For the bar graph, my estimates (in percent) are:

(a/X) _________ (b/X) _________ (c/X) _________ (d/X) _________.

THIRD PASS

For the pie chart, you are to judge what percent each wedge is of the wedge marked with an "X". For the bar graph, you are to judge what percent each bar is of the bar marked with an "X". The bar graph and pie chart are NOT related. **DO NOT** make any sort of measurements to obtain your estimates, just use a quick visual judgement. Record your results on the bottom of this page.

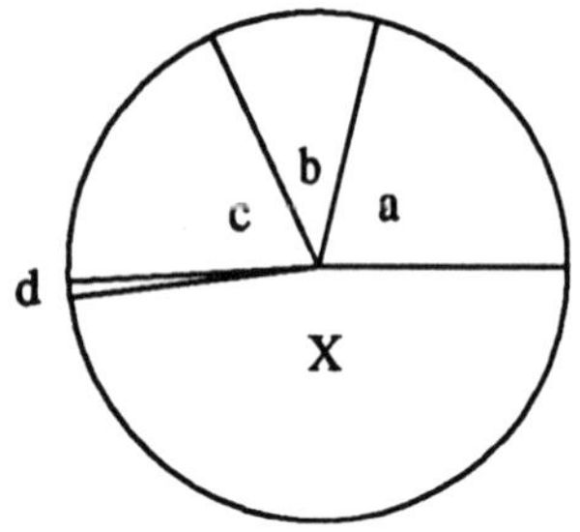

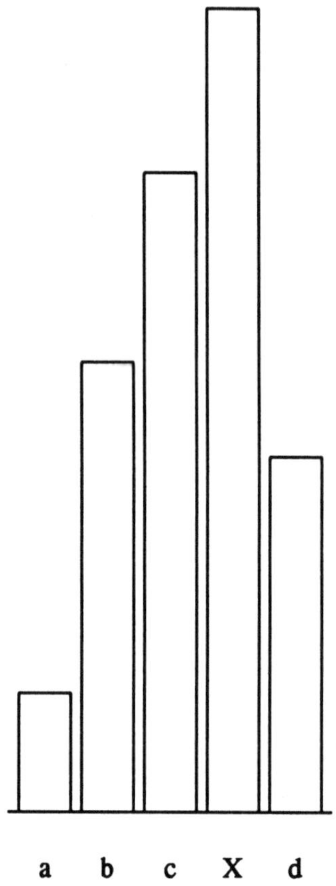

For the pie chart, my estimates (in percent) are:

(a/X) __________ (b/X) __________ (c/X) __________ (d/X) __________.

For the bar graph, my estimates (in percent) are:

(a/X) __________ (b/X) __________ (c/X) __________ (d/X) __________.

FOURTH PASS

For the pie chart, you are to judge what percent each wedge is of the wedge marked with an "X". For the bar graph, you are to judge what percent each bar is of the bar marked with an "X". The bar graph and pie chart are **NOT** related. **DO NOT** make any sort of measurements to obtain your estimates, just use a quick visual judgement. Record your results on the bottom of this page.

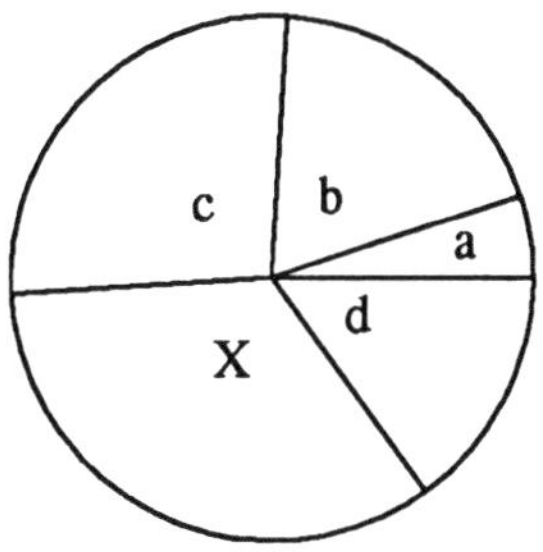

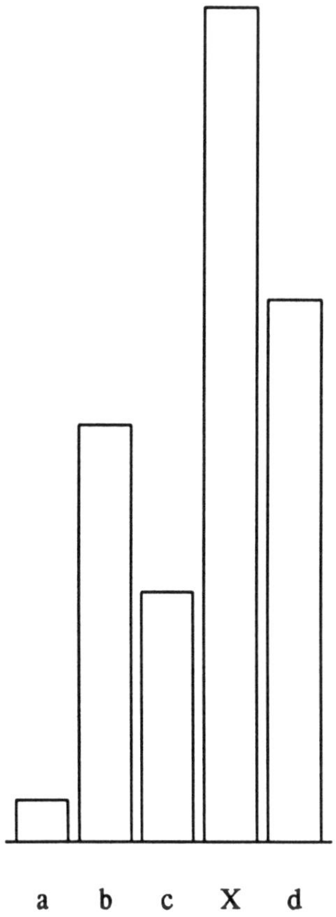

For the pie chart, my estimates (in percent) are:

(a/X) __________ (b/X) __________ (c/X) __________ (d/X) __________.

For the bar graph, my estimates (in percent) are:

(a/X) __________ (b/X) __________ (c/X) __________ (d/X) __________.

FIFTH PASS

For the pie chart, you are to judge what percent each wedge is of the wedge marked with an "X". For the bar graph, you are to judge what percent each bar is of the bar marked with an "X". The bar graph and pie chart are **NOT** related. **DO NOT** make any sort of measurements to obtain your estimates, just use a quick visual judgement. Record your results on the bottom of this page.

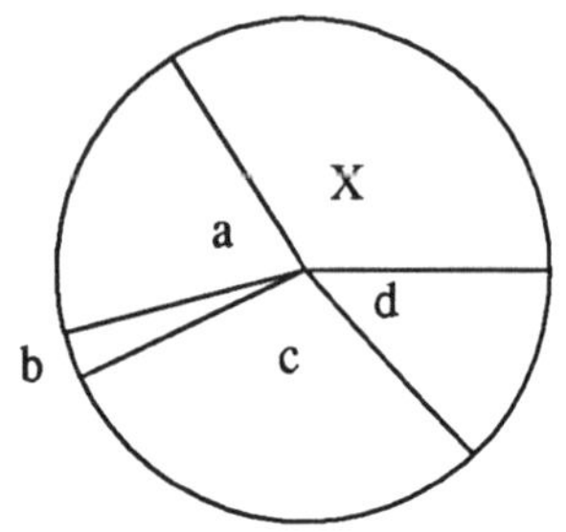

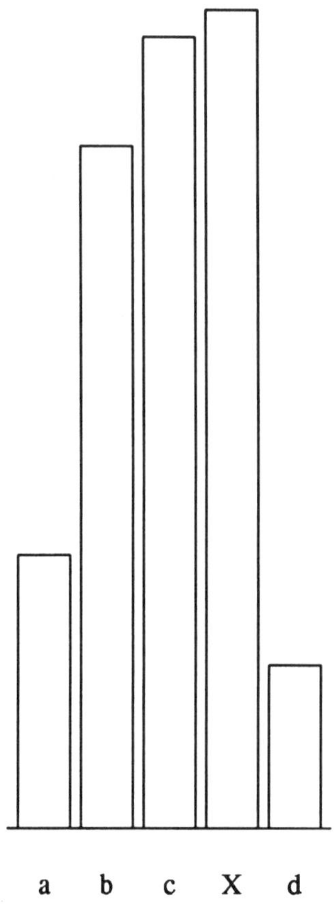

For the pie chart, my estimates (in percent) are:

(a/X) __________ (b/X) __________ (c/X) __________ (d/X) __________.

For the bar graph, my estimates (in percent) are:

(a/X) __________ (b/X) __________ (c/X) __________ (d/X) __________.

SIXTH PASS

For the pie chart, you are to judge what percent each wedge is of the wedge marked with an "X". For the bar graph, you are to judge what percent each bar is of the bar marked with an "X". The bar graph and pie chart are **NOT** related. **DO NOT** make any sort of measurements to obtain your estimates, just use a quick visual judgement. Record your results on the bottom of this page.

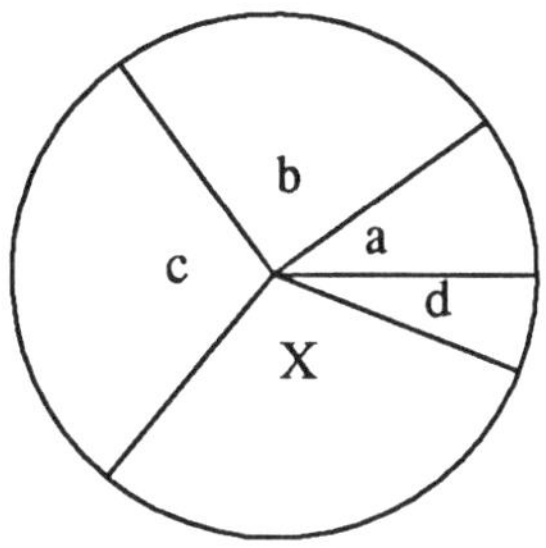

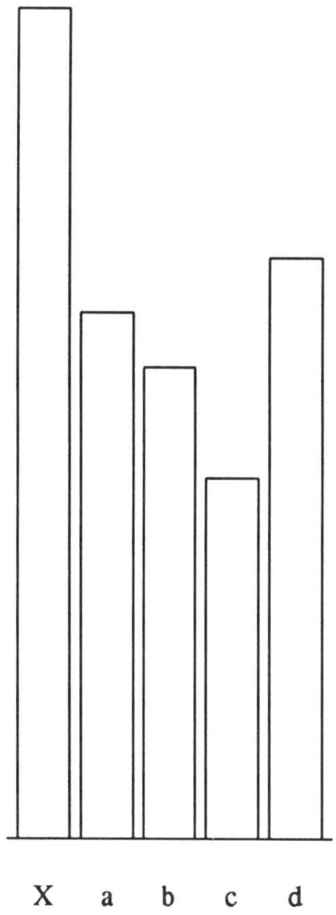

For the sixth pie chart, my estimates (in percent) are:

(a/X) __________ (b/X) __________ (c/X) __________ (d/X) __________.

For the sixth bar graph, my estimates (in percent) are:

(a/X) __________ (b/X) __________ (c/X) __________ (d/X) __________.

DATA SHEET **NAME________________________**

I am _______ MALE _______ FEMALE.

I am majoring in

_______ PHYSICAL SCIENCE or BIOLOGICAL SCIENCE or ENGINEERING

_______ HUMANITIES or SOCIAL SCIENCE.

If you had qualitative data to present, would you use a pie chart or a bar graph?

_______ PIE CHART

_______ BAR GRAPH

Record your estimates on this page.

first pie: (a/X) __________ (b/X) __________ (c/X) __________ (d/X) __________.

first bar: (a/X) __________ (b/X) __________ (c/X) __________ (d/X) __________.

second pie: (a/X) __________ (b/X) __________ (c/X) __________ (d/X) __________.

second bar: (a/X) __________ (b/X) __________ (c/X) __________ (d/X) __________.

third pie: (a/X) __________ (b/X) __________ (c/X) __________ (d/X) __________.

third bar: (a/X) __________ (b/X) __________ (c/X) __________ (d/X) __________.

fourth pie: (a/X) __________ (b/X) __________ (c/X) __________ (d/X) __________.

fourth bar: (a/X) __________ (b/X) __________ (c/X) __________ (d/X) __________.

fifth pie: (a/X) __________ (b/X) __________ (c/X) __________ (d/X) __________.

fifth bar: (a/X) __________ (b/X) __________ (c/X) __________ (d/X) __________.

sixth pie: (a/X) __________ (b/X) __________ (c/X) __________ (d/X) __________.

sixth bar: (a/X) __________ (b/X) __________ (c/X) __________ (d/X) __________.

Detach this page and hand it in.

11.6 ANALYSIS

The goal of this section, is to examine the class results. As you can imagine, these results are quite extensive. For example, for each measurement error, you know whether it is based on a bar or a wedge, the sex of the student, the academic major of the student and whether the student would prefer to use a pie chart or bar graph to illustrate qualitative data. Thus, for each student in the class, you collected 51 numbers. Due to the richness of the data, your instructor will provide you with two versions of the data to assist you in your analysis.

VERSION A

In this analysis, you will be examining the "typical" measurement error for your class. In particular, for each bar and each wedge you have a *distribution* of measurement errors. To simplify your analysis, rather than look at the entire distribution of measurement errors for each bar and each wedge, you will examine how the "typical" measurement error behaves. Your instructor will provide you with the median measurement error (over all students in the class) for each of the wedges and each of the bars, as well as the associated true percentages. You will use these data to answer the following questions.

11.1a. Draw a histogram for the median measurement errors made for the wedges. Draw a histogram for the median measurement errors made for the bars. Describe each of the histograms, remembering *shape, location and scale!* Compare the two histograms.

11.2a. Draw a box-plot for the median measurement errors made for the wedges. Draw a box-plot for the median measurement errors made for the bars. Describe each of the box-plots. Compare the two box-plots. How do the box-plots differ? How are they similar? What does this say about wedges and bars?

11.3a. For the wedge data, plot each of the median measurement errors versus the associated true percentage. Pass a straight line through this scatter plot. Does a straight line seem to provide a reasonable summary of the relationship between the median measurement errors for the wedges and the true percentages? If the answer is yes, what can you conclude about the relationship between the median measurement error and the true percentage? If the answer is yes, interpret what the slope means in this context and present a 95% confidence interval for the slope.

11.4a. Repeat question 11.3a for the bar data.

11.5a. Obtain a smooth regression curve for the wedge data using the running median smoother. (If you are using MINITAB, then you may use RSMOOTH. See the MINITAB APPENDIX for directions on how to use RSMOOTH.) Be sure that the pairs of numbers (true percentage, measurement error) are sorted with respect to the true percentages before passing the data through the smoother. Discuss the nature of the resulting regression curve. What can you conclude about the relationship between the median measurement errors and the true percentages in light of this curve?

11.6a. Repeat question 11.5a for the bar data. Compare the regression curves for the bar data and the wedge data. How do they differ? In what ways are the curves similar?

To answer the following question, your instructor will instruct you to produce a transformed version of your data. Your instructor will describe the rationale for this transformation.

11.7a. For the transformed version of your data, draw a histogram and a box-plot. Describe both plots. What do you conclude?

11.8a. Plot the transformed values versus the true percentages. Pass these data through the running median smoother. Describe the resulting regression curve. What do you conclude?

11.9a. Based on your responses to questions 11.1a-11.8a, which of the two tasks, the perception of angle or the perception of position along a common scale is more accurate?

VERSION B

In this section, you will examine the effect of sex, academic major and preference on the "typical" measurement error. Your instructor will provide you with the median measurement error for each of the wedges and each of the bars, broken down by sex, academic major and preference. In other words, for each bar and each wedge, you will be given the median measurement error for the men, for the women, for the physical science / biological science / engineering majors, for the humanities / social science majors, for those students who prefer to use bar graphs (bar lovers) and for those students who prefer to use pie charts (pie lovers). You will use these data to answer the following questions.

11.1b. Draw a box-plot for the median measurement errors made for the men's wedges. Draw a box-plot for the median measurement errors made for the women's wedges. Describe each of the box-plots. Compare the two box-plots. How do the box-plots differ? In what ways are they similar?

11.2b. Repeat question 11.1b for the bars.

11.3b. Compare the box-plot for the men's wedges to the box-plot for the men's bars. Compare the box-plot for the women's wedges to the box-plot for the women's bars. How do the box-plots differ? In what way are they similar?

11.4b. Draw a box-plot for the median measurement errors made for the physical science / biological science / engineering majors' wedges. Draw a box-plot for the median measurement errors made for the humanities / social science majors' wedges. Describe each of the box-plots. Compare the two box-plots. How do the box-plots differ? In what way are they similar?

11.5b. Repeat question 11.4b for the bars.

11.6b. Compare the box-plot for the physical science / biological science / engineering majors' wedges to the box-plot for the physical science / biological science / engineering majors' bars. Compare the box-plot for the humanities / social science majors' wedges to the box-plot for humanities / social science majors' bars.

11.7b. Draw a box-plot for the median measurement errors made for the pie lover's wedges. Draw a box-plot for the median measurement errors made for the bar lover's wedges. Describe each of the box-plots. Compare the two box-plots.

11.8b. Repeat question 11.7b for the bars.

11.9b. Compare the box-plot for the median measurement errors made for the pie lover's wedges to the box-plot for pie lover's bars. Compare the box-plot for the median measurement errors made for the bar lover's wedges to the box-plot for bar lover's bars.

11.10b. For the men's wedge data, plot each of the median measurement errors versus the associated true percent. For the women's wedge data, plot each of the median measurement errors versus the associated true percent. Obtain a smooth regression curve for the men's wedge data using the running median smoother. Obtain a smooth regression curve for the women's wedge data. (Again, if you have access to MINITAB, then you may use RSMOOTH.) Compare the two curves. How do they differ? How are they similar?

11.11b. Repeat question 11.10b for the bar data.

11.12b. Compare the two regression curves for the men (bar versus wedge). Compare the two regression curves for the women (bar versus wedge).

11.13b. For the physical science / biological science / engineering majors' wedge data, plot each of the median measurement errors versus the associated true percent. For the humanities / social science majors' wedge data, plot each of the median measurement errors versus the associated true percent. Obtain a smooth regression curve for the physical science / biological science / engineering majors' wedges. Obtain a smooth regression curve for the humanities / social science majors' wedges. Compare the two curves.

11.14b. Repeat question 11.13b for the bar data.

11.15b. Compare the two physical science / biological science / engineering majors' regression curves (bar versus wedge). Compare the two humanities / social science majors' regression curves (bar versus wedge).

11.16b. For the bar lover's wedge data, plot each of the median measurement errors versus the associated true percent. For the pie lover's wedge data, plot each of the median measurement errors versus the associated true percent. Obtain a smooth regression curve for the bar lover's wedge data using the running median smoother. Obtain a smooth regression curve for the pie lover's wedge data. (Again, if you have access to MINITAB, then you may use RSMOOTH.) Compare the two curves. How do they differ? How are they similar?

11.17b. Repeat question 11.16b for the bar data.

11.18b. Compare the two regression curves for the bar lovers (bar versus wedge). Compare the two regression curves for the pie lovers (bar versus wedge). How do the curves differ? How are they similar?

11.19b. Based on your responses to questions 11.1b-11.18b, can you conclude that men differ from women with regard to the ability to perceive angle or to perceive position on a common scale? Do physical science / biological science / engineering majors differ from humanities / social science majors with regard to the ability to perceive angle or to perceive position on a common scale? Do people who prefer bar graphs differ from people who prefer pie charts with regard to the ability to perceive angle or to perceive position on a common scale?

11.7 MINITAB APPENDIX

Given pairs of numbers $(x_1, y_1), (x_2, y_2) \ldots (x_n, y_n)$, MINITAB excels in obtaining the least-squares estimates of γ and β (and their *estimated* standard errors) for the model

$$y_i = \gamma + \beta x_i + \varepsilon_i \ .$$

If the y-values are stored in column C1 and the x-values are stored in column C2, then the command

> REGRess the y values in column C1 on the 1 column of data stored in C2 and store the residuals in column C3 and the predicted values in column C4

or more simply REGR C1 1 C2 C3 C4 , will return $\hat{\gamma}$, $\hat{\beta}$, their estimated standard errors and will store the *standardized* residuals in column C3 and the predicted values in column C4 for later examination. What are *standardized* residuals? We defined residuals to be

$$y_i - \hat{\gamma} - \hat{\beta}\, x_i$$

MINITAB *scales* these "regular" residuals to standard units, so that the residuals have an average of zero and an estimated standard error of one.

For example, if we give this command using the Millikan data (see Millikan, 1916), MINITAB returns

```
The regression equation is
C1 = - 6.83 + 0.0648 C2

Predictor        Coef          Stdev      t-ratio
Constant     -6.82585        0.05736      -119.00
C2          0.0648200      0.0006760        95.89

s = 0.03360      R-sq = 100.0%     R-sq(adj) =  99.0%

Analysis of Variance

SOURCE        DF          SS           MS
Regression     1      10.382       10.382
Error          4       0.005        0.001
Total          5      10.386
```

In this way, we see that the intercept is -6.826 with an *estimated* standard error of 0.057, while the slope is 0.0648 with an *estimated* standard error of 0.0007. MINITAB also reports several other quantities:

(1) t-ratio - This is simply the ratio of the estimate to its *estimated* standard error. A comparison of this ratio to the *t* graph on $n-2$ degrees of freedom can be used to test the null hypothesis that $\gamma = 0$ or that $\beta = 0$.

(2) s - This is the square root of the sum of the squared residuals divided by $(n-2)$.

(3) R-sq is the square of the correlation coefficient, i.e. r^2.

(4) R-sq(adj) - see Johnson and Bhattacharyya (1985).

(5) Analysis of Variance - see Johnson and Bhattacharyya (1985).

The command

PLOT C3 C4

yields a residual versus predicted value plot (see Johnson and Bhattacharyya, 1985), while the command sequence

NSCO C3 C5
PLOT C3 C5

yields a normal scores plot of the residuals (see Johnson and Bhattacharyya, 1985).

MINITAB does not have a command for directly computing the regression curve using *running medians*. However, it does have a fancier smoother RSMOOTH which can be used to filter a scatter plot. Given pairs of numbers (x_1, y_1), (x_2, y_2) ... (x_n, y_n), *with the y-values in column C1 and the x-values in column C2,* first give MINITAB the command

SORT the data in column C2 carrying along the corresponding values in column C1 and put them in column C2 and column C1 respectively.

or more simply, SORT C2 C1 C2 C1. The effect of this command is to sort the x-values, carrying along the corresponding y-values. Next, to obtain the y coordinates of the smooth curve, tell MINITAB to compute the

RSMOOTH of the y values in C1 putting the rough in column C3 and the smooth in column C4

or more simply RSMOOTH C1 C3 C4 . The y-values of the regression curve are in column C4 and can be plotted against the x-values in C1 to obtain the smooth regression curve.

11.8 REFERENCES

1. Alba, E. and Zartman, D. L. "Testing Outliers in Time Series: An Application to Remotely Sensed Temperatures in Cattle." Special Paper No. 130, Agricultural Experiment Station, New Mexico State University, 1979.

2. Cleveland, W. S. and McGill, R. "Graphical Perception: Theory, Experimentation, and Application to the Development of Graphical Methods." *Journal of the American Statistical Association, 79,* 531-554, 1984.

3. Cleveland, W. S. and McGill, R. "The Many Faces of a Scatterplot." *Journal of the American Statistical Association, 79,* 807-822, 1984.

4. Croxton, F. E. "Further Studies in the Graphic Use of Circles and Bars II: Some Additional Data." *Journal of the American Statistical Association, 22,* 36-39, 1927.

5. Croxton, F. E. and Stryker, R. E. "Bar Charts Versus Circle Diagrams." *Journal of the American Statistical Association, 22,* 473-482, 1927.

6. Croxton, F. E. and Stein, H. "Graphic Comparisons by Bars, Squares, Circles, and Cubes." *Journal of the American Statistical Association, 27,* 54-60, 1932.

7. Eells, W. C. "The Relative Merits of Circles and Bars for Representing Component Parts." *Journal of the American Statistical Association, 21,* 119-132, 1926.

8. Johnson, R. A., and Bhattacharyya, G. *Statistics: Principles and Methods,* John Wiley, New York, 1985.

9. Karsten, K. G. *Charts and Graphs,* Prentice Hall, New York, 1923.

10. Millikan, R. A. "A Direct Photoelectric Determination of Planck's 'h'." *Physical Review, 7,* 355-388, 1916.

11. Playfair, W. *Statistical Breviary: Shewing, on a Principle Entirely New, the Resources of Every State and Kingdom in Europe,* London, 1801.

12. Velleman, P. F. and Hoaglin, D. C. *Applications, Basics and Computing (ABC's) of Exploratory Data Analysis (EDA),* Duxbury Press, Boston, MA, 1981.

Appendix

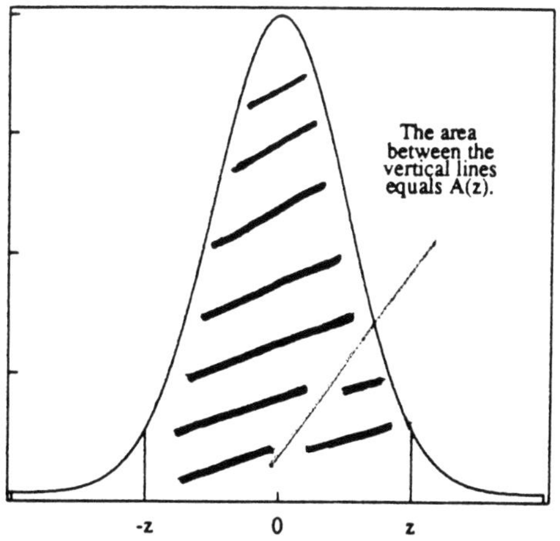

A Normal Table

z	A(z)	z	A(z)	z	A(z)	z	A(z)
0.00	0.0000	0.75	0.5467	1.50	0.8664	2.25	0.9756
0.05	0.0399	0.80	0.5763	1.55	0.8789	2.30	0.9786
0.10	0.0797	0.85	0.6047	1.60	0.8904	2.35	0.9812
0.15	0.1192	0.90	0.6319	1.65	0.9011	2.40	0.9836
0.20	0.1585	0.95	0.6579	1.70	0.9109	2.45	0.9857
0.25	0.1974	1.00	0.6827	1.75	0.9199	2.50	0.9876
0.30	0.2358	1.05	0.7063	1.80	0.9281	2.55	0.9892
0.35	0.2737	1.10	0.7287	1.85	0.9357	2.60	0.9907
0.40	0.3108	1.15	0.7499	1.90	0.9426	2.65	0.9920
0.45	0.3473	1.20	0.7699	1.95	0.9488	2.70	0.9931
0.50	0.3829	1.25	0.7887	2.00	0.9545	2.75	0.9940
0.55	0.4177	1.30	0.8064	2.05	0.9596	2.80	0.9949
0.60	0.4515	1.35	0.8230	2.10	0.9643	2.85	0.9956
0.65	0.4843	1.40	0.8385	2.15	0.9684	2.90	0.9963
0.70	0.5161	1.45	0.8529	2.20	0.9722	2.95	0.9968

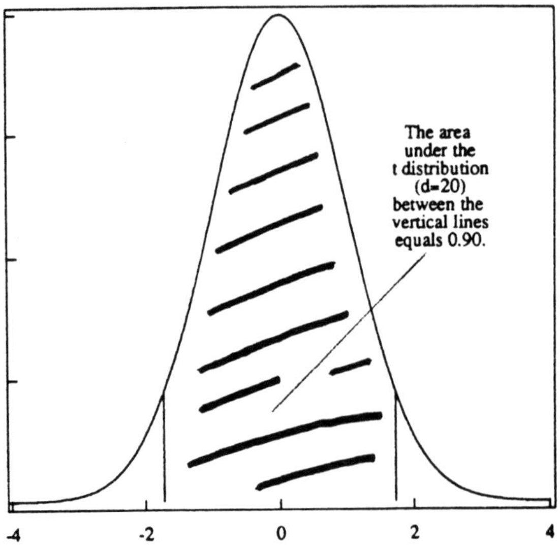

Areas Under Various t Distributions

df	0.80	0.90	0.95	0.98	0.99
1	3.078	6.314	12.706	31.821	63.657
2	1.886	2.920	4.303	6.965	9.925
3	1.638	2.353	3.182	4.541	5.841
4	1.533	2.132	2.776	3.747	4.604
5	1.476	2.015	2.571	3.356	4.032
6	1.440	1.943	2.447	3.143	3.707
7	1.415	1.895	2.365	2.998	3.499
8	1.397	1.860	2.306	2.896	3.355
9	1.383	1.833	2.262	2.821	3.250
10	1.372	1.812	2.228	2.764	3.169
11	1.363	1.796	2.201	2.718	3.106
12	1.356	1.782	2.179	2.681	3.055
13	1.350	1.771	2.160	2.650	3.012
14	1.345	1.761	2.145	2.624	2.977
15	1.341	1.753	2.131	2.602	2.947
16	1.337	1.746	2.120	2.583	2.921
17	1.333	1.740	2.110	2.567	2.898
18	1.330	1.734	2.101	2.552	2.878
19	1.328	1.729	2.093	2.539	2.861
20	1.325	1.725	2.086	2.528	2.845
21	1.323	1.721	2.080	2.518	2.831
22	1.321	1.717	2.074	2.508	2.819
23	1.319	1.714	2.069	2.500	2.807
24	1.318	1.711	2.064	2.492	2.797
25	1.316	1.708	2.060	2.485	2.787
30	1.310	1.697	2.042	2.457	2.750
40	1.303	1.684	2.021	2.423	2.704
50	1.299	1.676	2.009	2.403	2.678
60	1.296	1.671	2.000	2.390	2.660
80	1.292	1.664	1.990	2.374	2.639
Normal	1.282	1.645	1.960	2.326	2.576

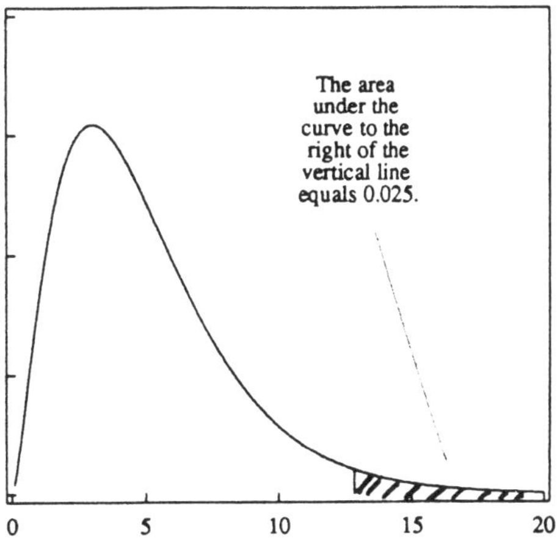

Areas to the Right Under Various χ^2 Distributions

df	0.95	0.90	0.70	0.50	0.10	0.05	0.025	0.01
1	0.004	0.02	0.15	0.45	2.71	3.84	5.02	6.64
2	0.10	0.21	0.71	1.39	4.60	5.99	7.38	9.21
3	0.35	0.58	1.42	2.37	6.25	7.82	9.35	11.34
4	0.71	1.06	2.20	3.36	7.78	9.49	11.14	13.28
5	1.14	1.61	3.00	4.35	9.24	11.07	12.83	15.09
6	1.64	2.20	3.83	5.35	10.65	12.59	14.45	16.81
7	2.17	2.83	4.67	6.35	12.02	14.07	16.01	18.48
8	2.73	3.49	5.53	7.34	13.36	15.51	17.53	20.09
9	3.33	4.17	6.39	8.34	14.68	16.92	19.02	21.67
10	3.94	4.86	7.27	9.34	15.99	18.31	20.48	23.21
11	4.58	5.58	8.15	10.34	17.28	19.68	21.92	24.73
12	5.23	6.30	9.03	11.34	18.55	21.03	23.34	26.22
13	5.89	7.04	9.93	12.34	19.81	22.36	24.74	27.69
14	6.57	7.79	10.82	13.34	21.06	23.69	26.12	29.14
15	7.26	8.55	11.72	14.34	22.31	25.00	27.49	30.58
16	7.96	9.31	12.62	15.34	23.54	26.30	28.85	32.00
17	8.67	10.09	13.53	16.34	24.77	27.59	30.19	33.41
18	9.39	10.87	14.44	17.34	25.99	28.87	31.53	34.81
19	10.12	11.65	15.35	18.34	27.20	30.14	32.85	36.19
20	10.85	12.44	16.27	19.34	28.41	31.41	34.17	37.57
21	11.59	13.24	17.18	20.34	29.62	32.67	35.48	38.93
22	12.34	14.04	18.10	21.34	30.81	33.92	36.78	40.29
23	13.09	14.85	19.02	22.34	32.01	35.17	38.08	41.64
24	13.85	15.66	19.94	23.34	33.20	36.42	39.36	42.98
25	14.61	16.47	20.87	24.34	34.38	37.65	40.65	44.31
30	18.49	20.60	25.51	29.34	40.26	43.77	46.98	50.89
40	26.51	29.05	34.87	39.34	51.81	55.76	59.34	63.69